URBAN POWER

AND SOCIAL WELFARE

URBAN POWER

AND SOCIAL WELFARE

Corporate Influence in an American City

RICHARD E. EDGAR

With a Foreword by Irving Louis Horowitz

SAGE Publications, Beverly Hills, California

For information address:

Sage Publications, Inc.
275 South Beverly Drive
Beverly Hills, California 90212

Printed in the United States of America

Library of Congress Catalog Card Number: 77–103480

Standard Book Number: 8039–0053–8

First Printing

So thou thyself from facts new facts shalt know,
And in such arguments shalt learn to creep
Into the secret lairs of hidden things
And thence drag forth the truth.

—Lucretius

FOREWORD

It is a distinct privilege to introduce Richard Edgar's excellent book, *Urban Power and Social Welfare: Corporate Influence in an American City*. It actually needs no prefatory sendoff to establish itself among that small handful of volumes that deserve to be treated seriously on the problem of how urban power becomes operational.

The importance of the book stems from the fact that this is one of the rare volumes which attempts a concrete investigation of a local power structure in a major American city—in this case St. Louis. Too much pluralistic writing takes place at local levels, such as New Haven and Oberlin, while too many power concentration approaches take into account primarily national and even international factors, leaving out of their reckoning more immediate variables related to community organization that may explain the functioning of power in American life. A chief merit of this volume is that it takes into account both "micro" and "macro" factors. In fact, Mr. Edgar's very unconcern for levels allows him to maneuver freely among his findings.

The chief findings of the book can be summed up directly. First, there does exist a power structure which is invisible to the popular imagination but quite visible to the serious investigators. Second, this power structure is, however, not sealed. It is subject to extreme pressures both from national, political and economic leadership and from local pressures from below. Third, power elites do not uniformly act in concert, at least not the power elite in St. Louis. The analogue of the family squabble is perhaps not inappropriate. As long as no external pressures are apparent, the normal psychological interplay creates considerable difference of opinion, while the customary tendencies of organizational life also accentuate such differences. Thus, it is only when "outsiders," whether they be local mass forces such as newly mobilized Negroes, or national elites such as Health, Education and Welfare officials, intervene or interrupt the normal flow of welfare funding, that the unity of the power structure becomes manifest. Mr. Edgar has succeeded in operationalizing a power elite theory for St. Louis without the customary spin-off into demoniacal pathologies or conspiratorial presumptions.

This volume was written from inside the welfare whale. It is not an

7

investigation of cities initiated by a team of scholars with a large grant and even larger hunch. Rather it is part and parcel of a scholar coming to terms with the nature of urban affairs and city power as it appeared to someone who spent more than a decade in the health and welfare establishments of Cincinnati and St. Louis. This very intimacy contrasts markedly with most writings on the subject of local power, and gives this book an empirical scope absent in most works of this genre.

Since this book was first written as a dissertation under my supervision, I see no reason to be unduly elaborate in a Foreword. Its major virtues will be clear enough, and they can be summed up by indicating how Mr. Edgar succeeds in taking an established body of theory in sociology and political science and linking it to a thoroughly untapped series of data imaginatively teased out of the local archives, newspapers, reports, and personal interviews. The ability of the author to weave a narrative around these materials is the basis of this work and needs no further rationalization.

No single work in urban affairs will settle the long-standing theoretical squabble between the power concentrationists and the power divisionists. However, works of this type have the supreme merit of pushing the "grand debate" one step further. Thus, if the argument cannot be resolved, it might actually be outflanked by a more inclusive kind of theory. In this sense, the utility of any theory is not necessarily confined to its truth value but may extend to its heuristic value. In this volume, Mr. Edgar has taken the two relatively undifferentiated frameworks of political elitism and political pluralism and has shown how genuinely fruitful they can be in conceptualizing the problems of poverty and power in American urban life.

Irving Louis Horowitz

Rutgers University
New Brunswick, New Jersey
March 28, 1969

INTRODUCTION

This book is a case study of who controls social policy in St. Louis. I am impressed by how major decisions are made by a few, and how little communication there is between these decision makers and the general public. This research differs from the work of Floyd Hunter (1953, 1959) and Robert Dahl (1961) by seeking the sources of influence and the control of policy in relations among agencies rather than in the general community at large. I conceptualize influence in terms of institutional position. The expansion of the federal government in the local community gives rise to questions of whether the control of social policy will be even more centralized, more pluralistic, or neither of these. Based upon the findings of this research, the conclusion is that private agency influentials control social policy, and their control is scarcely lessened by the rise of a public ideology.

Agency and policy as a doctoral dissertation topic was suggested to me by Professor Irving Louis Horowitz. His inspiration and my experience in a policy making setting gave the grist for the mill. Twelve years in health and welfare councils in Cincinnati and St. Louis plus a year with the St. Louis Community Renewal Program gave me an awareness of the interlocking of private and public agencies in policy making and its oligarchical character. Information on the private control of policy and its political organization comes from having been a participant observer in a series of events concerning the Health and Welfare Council of Metropolitan St. Louis' relations with other agencies. Although the participant observation was unplanned and unstructured, it did give me access to strategic information which was used in testing the hypotheses of this study. In addition, a framework for interpreting the facts was provided. Hence, it is from the vantage point of the Council that a number of observations are made about who really determines policy. Being cognizant of the limitations of the "insider" point of view, information has been supplemented by a qualitative study of policy making as reported in the two major dailies of the city for a period of more than a year. Finally, there were selected interviews to check certain observations. Once the crucial agencies in policy making are identified, tracing the sources and lines of influence is a matter of consulting various archives.

The guiding interest is who controls policy in St. Louis. It is evident from this research that business corporations are influential in the determination of the interagency organization of social policy. Operating through a network of organizations, they are able to mold the nature and development of social welfare. This state of affairs has important implications for freedom and equality. The curiosity sometimes expressed over whether or not there will continue to be private agencies in view of the expanding role of government may be overdrawn. It is more appropriate to ask if there can be public agencies which adequately represent the public interest. Is there really much of community consequence which private ideology, as expressed by the corporate power elite, does not influence? Public ideology as an ideal is often compromised by servants of power.

Business influentials rule because there is consensus that they should rule. It is consensus based on asymmetrical reciprocities since interdependence is skewed in the direction of the corporate elites. To depend on one's corporate bosses for food, clothing, shelter, and pensions for retirement is a simple pattern into which most men fit. It is one of superiors and inferiors, of imperative domination. Thus the business aristocrats derive substantial support from the populace who also embrace the private ideology. Consequently, corporate management stands legitimized and unimpeachable.

There is mobilization through organizations that oppose the large corporations. So far this has been relatively feeble resistance. There would be more dissensus if it were better understood who really governs the urban community.

The Welfare State is pseudo public ideology, and is better described as the Corporation State or what Lundberg calls the finpolity. The liberal's vision becomes the State which preserves the corporate domains and hence paradise lost. Phillip Hauser is probably correct in saying the Welfare State is no longer a pejorative term. Nevertheless, the Welfare State may be the Poorhouse State because it does not necessarily allocate economic wealth equitably, thus maintaining gross income inequities. Comprehensive planning in a rational model is urged by policy scientists as a sine qua non for the "good life," but the resources with which to plan and the undemocratic organization attained do not match the rhetoric.

In analyzing the data on St. Louis, it became apparent that the influence of governmental agencies was not as great as might be expected. The antipoverty programs serve as a case in point. Influence is limited in two fundamental ways: (a) the stipulations of the federal legislation on poverty are indicative of a commitment to the status quo; and (b) corporations besides being represented in the agency responsible for administration of the poverty program were also controlling its activities. The minimal standard of $3,000 as a dividing line between poverty and nonpoverty;

emphasis on work programs, make-work, and low pay point to an ideology consistent with the Protestant Ethic. Business influentials besides dictating functions of the local antipoverty agency, are giving it only token support. Although a public ideology appears to be increasing, corporations control the nature of it. However, in the process of protecting their own interests, there can be unanticipated consequences as social policy is enacted as law, leading at times to less control over policy by business influentials. Thus, the establishment is open to certain changes.

In looking further into corporate sources and lines of influence and control of policy, the conclusion reached is that one agency—Civic Progress, Inc.—is central in social policy decisions. Civic Progress does not confine itself to the narrow definitions of health and welfare, but rather ranges across a broad spectrum of problems and solutions akin to the Welfare State.

CONTENTS

TABLES

FIGURES

Part One

POLICY AND

AGENCY

Chapter 1

Control of Social Policy

The central problem of this book is who controls social policy in St. Louis. Social policy, inclusively defined, concerns what is done about all matters relating to the city's welfare. Economic allocation from this perspective is part of social policy. The welfare of the city and its economy are interdependent. Therefore, social policy refers to what is done, when, where, and for whom with respect to income, employment, housing, education, urban renewal, health and social services, recreation and the arts. Social policy pertains to slums, poverty, unemployment, alcoholism, crime, housing segregation, and anything else defined by the community as a problem.

Two questions guided the research: Do influential private agencies control interagency organization of social policy? What impact does the rise of public ideology have on influential private interagency control of social policy? [1]

To find out who decides social policy in St. Louis attention is given to the same kinds of questions asked by Robert Dahl (1961:7):

> Are inequalities in resources of influence "cumulative" or "noncumulative"? That is, are people who are better off in one resource also better off in others? In other words, does that way in which political resources are distributed encourage oligarchy or pluralism? How are important political decisions actually made? What kinds of people have the greatest influence on decisions? Are different kinds of decisions all made by the same people? From what strata of the community are the most influential people, the leaders, drawn? Do leaders tend to cohere in their policies and form a sort of ruling group, or do they tend to divide, conflict, and bargain? Is the pattern of leadership, in short, oligarchical or pluralistic?

Bollens and Schmandt (1965), Banfield (1965), Greer (1963), and Long (1962) follow Dahl by fostering the conclusion that St. Louis is pluralist. They do this by showing the fragmentation of the polity and by contending the voluntary sector is not coherently coordinated. Taking into account the prevailing mood of the times about metropolitan behemoths and social scientists racing about on chargers to slay dragons, one can understand the adventitiousness and functionality of their findings. Their work will be examined in the light of different facts and interpretations. However, first it is important to understand their arguments.

Bollens and Schmandt (1965:107) question private agencies having the cohesiveness to unify the community. After reviewing literature on elitism and pluralism they conclude that no one runs the metropolis (1965:196). Although their experience in metropolitan surveys includes other areas besides St. Louis, they also base their generalizations on this community. They do not find businessmen active in political affairs of the community.

> As Banfield found in Chicago, big business men are to be criticized less for interfering in public affairs than for failing to assume their civic responsibilities. Their role in public decisions is most frequently confined to rather passive membership on civic or governmental committees and to more active service in the private welfare sector of the community (1965:197).

Pictured are politically castrated businessmen:

> After relinquishing the political reins, commercial and industrial leaders became content to influence the conduct of government indirectly through various citizen groups and reform leagues. More importantly, they begin to play predominant roles in the private welfare sector of the community. Service on boards such as the Community Chest, Red Cross and hospitals became a substitute for political involvement. Activity of this type served several purposes for the economic notable. It furnished him with a means of satisfying his traditional sense of civic obligation without becoming immersed in local politics. It provided him with a highly prestigeful and noncontroversial role in civic affairs. And finally, it enabled him to retain certain responsibilities within control of the private sector of the community that otherwise would have to be assumed by government (1965:199).

To further clinch their argument for the absence of a corporate elite influence in political affairs, they comment on change in the control of corporations from family to nonfamily operated, and with the executives often from other communities. High mobility of these executives, they point out, discourages personal commitment; and when they do participate, they want it to be noncontroversial. Finally, although there is consultation in urban redevelopment programs, the relationships are usually with the high level politicians such as the mayor or city manager, and corporate tie-in with local political action is peripheral (1965:199–200).

Bollens and Schmandt attack the "mythology" of "Big Labor," "Big Business," and "Big Government" (1965:201). If there is much influence, they contend, it must be happening behind the scenes.

> Seldom do clashes over issues of noneconomic character stir the community waters. In civic causes, more often than not, trade union leaders will be found in the same camp with the business notables, assisting a chest or hospital drive, endorsing a bond issue for public improvements, working to establish a cultural center, or supporting an urban redevelopment project (1965:202).

Therefore, they conclude there is no triadic struggle.

Contrary to what Dahl found in New Haven and metro reformers in St. Louis, the facts presented in this research point to an oligarchical pattern of leadership, and thus back up Hunter's (1953) basic thesis, and are not inconsistent with other studies reporting elite networks even though they are on a national level (Mills, 1956; Domhoff, 1967; and Lundberg, 1968). An ideological gap exists between political science and sociology in the measures used to evaluate who rules urban communities. Dahl overlooked the sociological structuring of influence and control through the large corporations which, interlocked by reciprocities, have much to gain from maintaining close relationships with legislative and executive branches of government as well as nongovernmental agencies. Moreover, the coercive power of wealth and prestige to dominate imperatively is hidden by Dahl's blinders.

Amongst the oligarchical pattern is a pluralistic one. Thus, if one follows the warp and woof of pluralistic designs ignoring the elitist, the conclusions to be drawn about policy making are predictable. Playing upon the indicators of the electorate, governmental officials, and the multitude of quasi-public and voluntary agencies, and the thousands of businesses, the pattern is ambiguously pluralistic, and power is apparently diffused. However, by studying organizational networks of power based on social policy decision making, nonpluralistic designs are evident.

Elite influences are manifested in lines of power between:

(1) The largest locally based business corporations in the form of interlocking directorates.

(2) The largest locally based business corporations and lesser business corporations in terms of many social policy considerations (e.g. philanthropic and charitable fund raising) and simply because the larger corporations are significant purchasers from lesser corporations.

(3) A voluntary agency (Civic Progress, Inc. which is composed of

the largest locally based business corporations) and public, quasi-public, and other voluntary agencies.

(4) The same largest locally based business corporations and public, quasi-public, and voluntary agencies (e.g. McDonnell Douglas and its relations with the Mayor's Office, Model City Agency and United Fund).

This interorganization of business, political, and educational elites dominates different sectors of social policy: employment, income, urban renewal, antipoverty programs, public assistance, medicare, economic planning, physical planning, social planning, educational expansions of universities, arts and symphony, charitable and philanthropic fund raising. It is apparent that policy decisions pertaining to these various sectors are made by the same network of agencies.

The major theme of this book is that interorganization of policy making is controlled by influential nongovernmental agencies linked in a business dominated gravitational field (Mills, 1956: Ch. 6 and 7; also see Domhoff, 1967: Ch. 2). Nevertheless, the control is not totalistic or monolithic. The power structure of St. Louis is composed of different divisions of interest, at times in disharmony. Blacks, elected or appointed officials, agency professionals, social scientists, students, lower echelon executives represent different interests which may be in conflict among themselves or with business executives. Moreover, levels of interest vertically cut the community. Federal, state, and regional offices and officials interact and thus add to the profusion of interests. However, the many ripples on the surface of a river, its variable currents and different depths do not make it less a river. Likewise the pluralism of social policy actors does not preclude the existence of local pyramidal or hierarchical structure of organizational influence interpenetrating other levels of the body politic. A rent strike at a public housing project such as Pruitt-Igoe in St. Louis involved the Mayor's Office, local and federal housing officials, the local and national poverty agencies, and the Urban League which is interlaced at local and national levels with large scale business interests. Tensions and conflicts stemming from the organizational roles are indicative of an unequal contest between rulers and ruled. The organizational relations differ in terms of power or the potential to use force. The greater potential rests with the organizations linked to the large corporations and to government. The latter serves a brokerage function for real community managers who, though not elected by the people, are more powerful than the aristocracy of the eighteenth century simply because they have the resources, the numbers and the organization to be more influential actors. In an historical sense, big corporations have merged the functions of the aristocracy and the bourgeoisie. They are the new aristocracy. These giant private governments stand at a pivotal point in history where some like General Motors, larger than many public governments, have worldwide dominions menacing the nation state itself. The surge of conglomerate mergers during the 1960s, and the worldwide conquests by such private governments, in the form of multinational companies, throw into distinct relief the possibility of the end of public government (*Time,* March 7, 1969; *New York Times,* March 8, 1969 and March 13, 1969).

Inordinate power comes not only from the magnitude of the resources, numbers, and the organizational expertise but the interlocking in mergers, collusion, and through voluntary organizations like Civic Progress, Inc. in St. Louis or the Area Progress Council in Dayton. Not only do chief executive officers sit on each other's boards of directors, which foster reciprocal profiteering, but they work through such groups to further community and national social policy which are most importantly related to their own profits. Such corporations as McDonnell Douglas, Anheuser-Busch, Monsanto, Ralston Purina and others are preeminent simply because of oligopolistic concentration of power resources and because other centers in a community's power structure are dependent upon them for jobs, pensions, and profits for the continuance of life itself. Such social scientists as J. L. Talmon (1961) apparently repress this social fact, glaringly a part of the politics of realism, in celebrating liberal democracy.

Community pluralism, as well as pluralism in a vertical sense, is functional for operation of such organizational power concentrates. Opposing power accumulations are uncommon and unlikely events. The multitude of private interests is power erasing. People fear big government more than big corporations according to a Gallup Poll in 1968. During Eisenhower's term it was big labor that people perceived as constituting the greatest threat. The pluralism is functional for power ambiguities and misperceptions.

The general public is less interested in participatory governance than in consumption; corporations give material joys and governments take them away (Marcuse, 1966: Ch. 1).

CONCLUSIONS

Although apparently pluralistic, major social policy is really decided by influential private agencies in St. Louis. The social structure of private control and lines of power are through large locally based corporations which have a voluntary organization, Civic Progress, Inc., representing and furthering their interests and holdings. The unresolved conflict between private and public agencies for determination of public policy has significant implications for freedom and equality. The survival of a viable polity is in doubt because of the disproportionate power of these corporations.

Chapter 2

Ideology, the Corporation,

and Social Policy

What influential agencies do about urban problems reflects the ideology of such agencies. Policy is the more or less explicit manifestation of ideology. Idealizations buttressed by political and economic rationalizations which guide elective and nonelective influentials to establish models from which specific programs of action are derived and become institutionalized in the community is ideology (Horowitz, 1964a:50). Idealizations are the collective representations of activities, ambitions, and aspirations which function to justify authority in terms of economic and political theory. Rationalizations are the reasons given for idealizations and legitimate the organization of social policy desired or attained. Principles of action are the specific procedures for achieving or maintaining the kind of social order which is idealized, and operate to muster support for influentials. Ideology may be dichotomized, as ideal types, into private and public.

PRIVATE IDEOLOGY

Private ideology is oligarchical control of a social system. It is motivated by the ethics of private responsibility, hard work, frugality, and economic profit making (Weber, 1958). Its central agency is the large business corporation which Lundberg (1968:255) labels a "finpolity"—a financial state:

These are not businesses at all as the term has been historically understood. They are clearly more like governments, or government departments, and would be more aptly termed "finpolities." Their influence on formal government, direct and indirect, conscious and unconscious, is enormous. Their influence, indeed, is so often pre-emptory that it might better be described as in the nature of quasi-decretal. For such entities, through agents, often tell governments, in secret conference (the United States government included) what they must do and what they cannot do. That, I submit, is power. And, if governments fail to comply, at the very least they will lose the considerable cooperative power of the "finpolities."

The business corporation is a private government responsible to the largest shareholders. Lundberg explains why such companies can operate with virtual autonomy from most stockholders (1968:435–436). Members of the board of directors know who speaks for the big stockholders where the actual power lies.

Despite all the devotion to voting in corporations, the process is hardly democratic because the vote, in any showdown, is by shares of stock, not by individuals. All the thousands of rag-tag stockholders in the Ford Motor Company could not outvote the Ford family. The situation is absolutely or effectively the same in every company, which means that a very large and paramountly vital part of internal American affairs is under essentially autocratic rule, as in Russia.

The sine qua non for the corporation is profit. The profit criterion is rationalized by saying that making money is beneficial to the community. It can be likened to the United Fund slogan: "Everyone gives and everyone benefits." Hence, what is good for the corporation is good for the public. Frederic Peirce, President of General American Life Insurance (a St. Louis Civic Progress company) in a recent speech ("The Ghetto: A Business Interest," extracts reported in the *St. Louis Post-Dispatch,* October 13, 1968) noted that forty-three years ago Calvin Coolidge said "The business of the United States is business," and that Theodore Sorensen today suggests that "The business of business is America." This sums up the ideological take-over of America by business. Profits depend upon a ghetto that does not explode in the face of downtown interests in St. Louis, Chicago, Cincinnati, Dayton or other large central cities. It is further reasoned that a healthy and happy people can work harder for the corporation; contribute more toward community betterment and thus to the maximization of profits.

Such enlightened "finpolitanship" has not been exactly common currency in the land. It came at the end of a Molotov cocktail rather than through spontaneous magnanimity. "The public be damned" attitude is much more sophisticated today. Hunger in the ghettos, death at an early age, and racism belie the public relations image of publicly responsible corporations and chief executives. The shame of our cities—our way of life—appears also in various forms of pollution. For example, much has been done to eliminate air pollution in St. Louis. Its industrialists, however, continued burning soft coal, not through ignorance or insensitivity, but because to use hard coal would have reduced the competitive strength of their businesses in the area. Therefore, from the standpoint of private ideology, what helps those companies, benefits those who derive their livelihood either directly or indirectly from industries polluting air which St. Louisans breathe!

Corporations control the three key types of welfare: (a) corporation centered welfare, (b) private community welfare, and (c) public community welfare.

Employees of corporations look to the corporation for life's necessities and luxuries. Corporation centered welfare is primary welfare; it is paid from corporate earnings. The corporate welfare bonanza comes in the form of wages, salaries, bonuses, pensions, insurance benefits, health services, and other ways that the faithful can be rewarded.

Union welfare, a type of corporate welfare, depends on contributions of employees whose money comes from the corporation. Nothing less than pathos is evoked by the following quotations (Chinoy, 1961:158) which indicate what three-quarters of a century wrought for the American labor movement:

> Class consciousness in American society has had its ups and downs, emerging strongly in some periods, as in the early 1880s and the 1930s, years of depression and hardship, and subsiding at other times, as at present. In 1881 the assembled delegates to the first convention of the American Federation of Labor adopted a Declaration of Principles whose Preamble began: "A struggle is going on in the nations of the civilized world between the oppressors and the oppressed of all countries, a struggle between capital and labor, which must grow in intensity from year to year and work disastrous results to the toiling millions of all nations if not combined for mutual protection and benefits." The Preamble to the Constitution adopted in 1955 by the United American Federation of Labor-Congress of Industrial Organizations begins: "The establishment of this federation . . . is an expression of the hopes and aspirations of the working people of America. We seek the fulfillment of these hopes and aspirations through democratic processes within the framework of our constitutional government and consistent with our institutions and traditions."

Corporation centered welfare also includes social policy formulation and implementation like providing training programs for hard core unem-

ployed for which corporations are either reimbursed or do at a profit. For
example, the Delta Corp. in St. Louis operates a woman's job corps center
as a money making venture. A corporate welfare program may entail
building a new city as the Rouse Corporation did in Columbia, Maryland.
Then it persuaded Antioch to set up a college there.[2] Thus Antioch becomes
one of many private institutions being used to legitimate social policies of
inequality and profit making (Ridgeway, 1969: Ch. 3). In the above-
mentioned examples, as with the payment of wages to employees, the
welfare function is definitely interrelated with the business of making
profits.

Private community welfare, the second basic form, is still under the
dominions of corporate control through sources of funding, personnel, and
organizational resources. It is to be distinguished from corporation centered
welfare in that it creates and functions through voluntary agencies representing different interests. For example, Civic Progress, Inc. is a voluntary
organization made up of chief executive officers of local firms. Downtown,
Inc. and the Chamber of Commerce are similar organizations but subordinate in status, prestige, and role to Civic Progress. Similarly, the United
Fund, a voluntary agency, although dominated by Civic Progress influentials, includes various community groups. The Health and Welfare Council
(HWC) of Metropolitan St. Louis has member agencies both private and
public. Businesses, as such, do not belong to HWC but they do control it
through the United Fund which raises most of the money needed for this
planning organization. Important policy issues are cleared through business
influentials. Ordinarily, a committee is set up, with such an elite as chairman. Boy Scouts, Goodwill, YMCA are some of the many kinds of
agencies making up private community welfare but thoroughly controlled
by finpolities.

Big individual giving has been replaced in large measure by bureau-
cratic, draconic corporate solicitation. The largest corporations coerce the
smaller. Both corporate and union cadre cajole employees. Corporate
philanthropy legitimizes private ideology and is required of lesser companies, which can be imperatively coordinated because they sell merchan-
dise to larger companies.

The tertiary type is residual welfare under public auspices; corporate
controlled simply because of being part of the overall finpolity welfare system which regulates how much money is available for everyone's welfare.
In addition to corporate influence, the Protestant Ethic plays no small part
in keeping this type of welfare minimal. It is differentiated from the
preceding two types because it depends on public funds or tax dollars.
Corporation centered and private community welfare use but do not depend on public funds. Before communities were so beleaguered by black
discontent, public agencies were definitely third rate outfits—social work

for the poor. Corporation and private community welfare prided themselves in serving the entire community.

With public housing, urban renewal, War on Poverty, Medicaid, Model Cities, and other federally funded programs the public sector became more politically significant, if only slightly upgraded. Although corporate influentials gained directly from these programs (land clearance for industrial development, riverfront redevelopment, etc.) this by no means removed it from the residual welfare category. Money for public social policy, because it has time cutoffs and matching requirements, does not have the certitude that it has in the other two sectors.

In the more established publicly financed welfare, such as public assistance or "the welfare," it is "low living on the hog." It is bare-boned welfare standing in sharp contrast to more generous corporate welfare programs. Public community welfare is thoroughly identified as the Poorhouse State. Besides saving corporations money and of course tax money—keeping employees from being more unhappy with even higher deductions—it follows the idea that if public welfare were made too attractive people would not want to work for a living. Moreover, the residual role underscores that public agency is inferior to private agency or corporation.

The War on Poverty and the Model City program represent a more recent, yet precariously funded effort, to provide services for those heretofore neglected by corporate centered and private community welfare. In spite of the Office of Economic Opportunity (OEO) and Housing and Urban Development creating frictions by vertical interorganizational penetration in local communities, corporate monitoring through officerships and board memberships in such federally supported programs is omnipresent. Today, OEO programs are partially emasculated and overshadowed with a new people's circus, the Model City agency. The latter, too, leans on finpolities, for that is where real community power lies.

If one still has doubts about the substance of invidious comparisons between private and public welfare let him unobtrusively observe. For example, visit waiting rooms and offices of the three basic types of welfare organizations. The inferiority of what the tax dollar gets must be made crystal clear. Otherwise, people might come to prefer public to private ways.

The Protestant Ethic is also manifested through hard work. In spite of myths about leisure life, corporate leaders are not often idle. Just getting by is not enough. Initiative and innovativeness often demand total commitment. Like proprietors of old, chief executives and upwardly mobile executives are extolled for long hours in the office and in travel. In short, the business of their lives is business, just as "the business of the United States is business."

Conspicuous waste, although uncommon for the corporation, is business as usual in goods produced for the public (Packard, 1963: Ch. 4–12).

Planned obsolescence is the order of the day: a new model every year. It is a race with the junk yard as Willy Loman said in *Death of a Salesman*. What outlasts the final installment payment may truly be called "durable goods." The capacity to produce greatly outdistances the ability to consume where there is inequitable distribution of income. Therefore, stimulation of false needs keeps profits high and the economy strong. However, corporations are under pressure from large stockholders to manage their own expenditures with prudence. Poshness of executive offices is not waste but indicative of successfulness, insuring satisfied executives, and smooth functioning systems. It is prestigious to have spacious, expensively furnished offices for the higher executives. Both waste for consumers and for corporations is only wasteful in the sense of not being in the public interest. Frugality, on the whole, applies to activities of the corporation except where large expenditures are made in anticipation of higher profits. However, where waste can be used for some intended gain and covered in part by tax deductions, such waste is spread over all tax payers. In this sense, waste ceases to be a private responsibility and like pollution of air and water, follows a general welfare principle. That is, waste should be absorbed by the general public according to this private ideology, when possible, because this enhances the corporation. The criterion of frugality would seem to be met—even though achieved at times by plundering the public purse. As titans were, so corporations are legitimated in such activities through legislative and juridical systems (Stern, 1965).

Welfare in the real public interest is constrained by "privatizing" the governmental sector so that public programs often serve the ends of nongovernmental agencies in maintaining the status quo and adding to their own aggrandizement. Because private interests are so intertwined with governmental and nongovernmental agencies, it is difficult to conceive what is meant by public ideology.

PUBLIC IDEOLOGY

Public ideology rejects the profit motive, opting for economic equality, upward rather than downward mobilization of action, and denationalization. Both State and individual act in the interest of the many. However, deviance is recognized as vital to the public interest because the former is a well spring of change.

Repudiated, then, is private ideology and of course its arch organizational type: the finpolity. Envisioned is drastic redistribution of power. How to do this with a minimum of violence is the major question of this century.

Discarded is the notion that any segment should have disproportionate power so that the public interest would be frustrated. Public ideology allows for free association outside the polity but it is not advocacy of pluralism characterized by people "doing their own thing." Egoistic actions should not be permitted to destroy the possibilities for community and life itself. This is not a denial of individuality except where individuality, group or individual, endangers the welfare of all. Underlying the concept of public ideology is the assumption that certain actions are beneficial, neutral, and harmful. It is the basic task of science to assess the public interest. Science is today engaged in such endeavor, but too much of what is being done serves private ideology. It is the hallmark of a primitive society that the welfare system can be defined in terms expressed in the preceding section.

Public ideology means economic equalization of wealth. Economic disparities are ipso facto social and physical deprivation creating a State and society antithetical to freedom as well as to equality. Gross inequalities in wealth still exist in the United States (Lundberg, 1968: Ch. 1). It is not enough to eliminate poverty by provision of minimum standards. Adequacy is the goal. The U.S. is not enough, we should strive for income sufficiency throughout the world. Such income escalation is unimaginable without maximizing automation. Essential for the achievement of these objectives is the elimination of private ideology which curtails production for favorable profits and wages. Moreover, capital expansion is hindered because of commitment to the profit system itself.

Public ideology is not totalitarian democracy because of the insistence on participatory democracy. Such democracy is ineffective now because corporate structures, which make important social policy decisions, determine the nature of the welfare system. Black, working class, and student revolts are signs of upward mobilizations. Talmon (1961:1–2) in discussing totalitarian democracies of the left fails to conceive of the actual workings of liberal democracies with top-down problem solving through finpolities which enslave mankind in a work process. Man can never be really free until he has the option of either engaging or not, in productive labor (Marcuse, 1966). Talmon's notion of nontotalitarian democracy implies the liberty to shackle billions to a dehumanizing process. It is outmoded by technology monopolized in the interests of a few hundred finpolitan corporations. Such mass slavery, although acceptable to most people is actually no longer necessary. True freedom is both political and economic. Political freedom in liberal democracies means most people are forced to work in order to survive. This is delusive freedom.

The foregoing is not stated to justify totalitarian democracy. The so called democratic centralism is no more valid than is a liberal democracy which is de facto totalitarian (e.g. finpolities) politically as well as economically. Such totalitarianism of the center, although neither state collectivist nor dominated by reactionary private enterprise, does not legitimate its status if it, too, deprives liberty in significant ways.

(c) oligarchical decision making, and (d) commitment to social policies of inequality. Three types of welfare, underwritten by the Protestant Ethic, are described.

Opposed to private ideology is public ideology, which negates the foregoing principles and opts for maximum deviance possible, congruent with achieving economic equality.

Policy sciences play a crucial part in making policy rational. In so doing they are ideologically involved which implies potential limitations on their freedom to criticize. In rationalizing policy, they may unwittingly legitimate private ideology.

Findings in the chapters which follow do not justify the contention that public ideology is significantly increasing. In the early stages of this research it was supposed that as public ideology increases, the control of social policy by influential private agencies decreases. It was hypothesized that the increased influences from federal, regional, and local sources (due to the power of federal funds) would lessen the private interagency control of social policy. Substantial increase in public influence has been aborted. Problems of cities, whether physical decay or social deprivation, are indeed national in scope. These problems appear too overwhelming for local corporate elites to ignore or cope with by themselves. However, instead of philanthropy or corporate influence declining, or lower status leadership interpenetrating the higher policy circles, and their being less dependent on the business influentials to raise money, the status quo seems essentially unchanged. Moreover, private ideological influence structure waxes strong, and paradoxically, polities wither in the dazzling course of finpolities.

Part Two

TENSIONS AND

CONTRADICTIONS IN

POWER STRUCTURE

Chapter 3

The Fall of Polity

and Rise of Finpolity

The fifties marked the failure of civic influentials to secure governmental reforms in St. Louis and the rise of finpolity social policy. Polity reforms were unsuccessful, and it was the organization of Civic Progress which brought some degree of coherency in a fragmented urban scene. Elitist factions combined in an attempt to meet the challenge of "progress or decay."

Despite elitism in St. Louis, there were definite tensions and conflicts among organizational actors: finpolitan interests, the blacks, and organized labor over governmental reforms. The disproportionate concentration of power in finpolitan St. Louis did not mean the absence of the usual urban problems: racism, housing, crime, poverty, too many governments, economic decline of the city, and municipal bankruptcy. In fact, social policies of inequality create these kinds of problems and keep power concentrations off balance by pressure from different elements of the power structure. Greater awareness of where this power lies, how it is used and distributed would increase conflict even more.

Corporation directed welfare, and its subordinate community forms, bogged down in the 1960s because of being unable to cope with the black revolution against racism. Business leaders, with financial holdings in jeopardy, were not caught unawares. They could see the writing on the wall in the 1950s long before black power was scrawled there. Moreover, it was not just a St. Louis problem. Major northern metropolitan areas had the same basic shifts in population: Negroes moving into their central cities and whites moving to outlying suburbs and counties. In the City of St. Louis, between 1940 and 1950, the Negro population increased from 109,000 to 154,000. By 1960 it had added another 60,000. During the same period, the whites declined from 707,000 to 534,000 (U.S. Bureau of the Census, 1940: Table 21; 1952:258; 1960a: Table 96).

The political impact of the blacks was felt during World War II years with the migrants coming primarily from Mississippi and Arkansas to fill defense jobs and the generally expanding labor market. The Negroes voted Democratic and St. Louis had its last Republican mayor in 1949.

For St. Louis, the city-county split started back in 1876, when, by constitutional amendment, the City of St. Louis divorced itself from St. Louis County. In the course of almost a century, nearly a hundred municipalities grew up around the city. St. Louis County, with most of these municipalities and unincorporated land, was farms and wooded countryside for the most part then. Now, in 1969, the county population is larger than the city.

The problem came to be both visually and statistically obvious. The county was more white and more affluent than the city. In population terms, the county's Negroes were stationary while the whites increased spectacularly (U.S. Bureau of the Census, 1960a):

	1940	1950	1960
Negro	12,000	17,000	19,000
White	262,000	389,000	684,000

The median family income level of the county in 1959 was $7,500; for the Negro, $3,600. For the Negro in the city, it was $3,700 (U.S. Bureau of the Census, 1960b).

The Charter reform fight of 1950 was probably the turning point. Banfield (1965:130) states:

> The reformers wanted to create a broader taxing authority, thus increasing revenue for the support of municipal services, and they wanted to strike a blow at machine politics. They proposed centralizing authority under the mayor, eliminating county office patronage, putting county offices under the mayor's fiscal control, and levying an earnings tax. The daily newspapers and the businessmen quickly endorsed these proposals. All but one of the ward politicians opposed them, however, as did the Teamsters and most of the Negro leaders. The Charter reform of 1950 was beaten.

By the early 1950s it became apparent to Mayor Darst that the city's problems had considerably worsened with the ever growing split between city and county. In 1953 he formed Civic Progress, and reflecting on its accomplishment, he said that St. Louis was just not getting anything done until these men took hold. The origins of Civic Progress, according to Banfield, came from concern over the changing composition of the electorate because of the sizeable in-migration of Negroes after World War II (Banfield, 1965:129). Through Civic Progress' resources an earnings tax passed in 1954; the following year $110 million in public improvement bonds for the city and a $39 million bond issue for the county were approved. The Community Chest was also reorganized into the United Fund.

Brimming with success from these and other ventures Civic Progress succeeded in electing freeholders to draft a new charter in 1956. The proposed revision provided for electing half the aldermen at large. Banfield says this was to reduce the influence of special interests. The powerful Teamsters Local 688 interpreted the proposal as aimed at depriving Negroes of political power and rallied labor and the blacks again. "Negroes (who would lose representation), organized labor, some ward leaders, neighborhood business and the community press [he is referring to neighborhood papers] were all against it" (Banfield, 1965:130–131). The proposed charter revisions were defeated in the 1957 election. This suggests some openness in the community power structure; chief executive officers do not win every time. It is not, however, the isolated victories or defeats that give rise to dismay but the swamping, in a statistical sense, that permits generalizations about modal tendencies.

METRO REFORMERS

But the community stage was actually being set for a subsequent attempt to merge city and county. Simultaneously with the move to change the Charter in 1956, Washington University and St. Louis University had received a $250,000 grant from the Ford Foundation and $50,000 from the McDonnell Aircraft Corp. Charity Trust to study the problems standing in the way of metropolitan government. Scott Greer (1963), chief sociologist for the St. Louis Metropolitan Survey, used the events, surrounding the failure of the proposed District Plan to be approved by voters in 1959, to support a pluralist hypothesis concerning power relations. He omitted key facts, and accordingly, crucial interpretations, which would shine a more elitist light on the topology of happenings. It is well to go into the background of this matter because: (a) the St. Louis Metropolitan Survey was not a unique event and (b) others involved in this survey came to pluralist conclusions about power configurations. These kinds of surveys occupied social scientists in many metropolitan areas during the late 1950s. It was the social science elan vital of a decade; just as the War on Poverty was the Zietgeist of the 1960s.

Climbing above the scatter of events cluttering the fifties, taking a perspective that finpolities rule and that they are interrelated not only at local but intercity and national levels, one is awestruck by how well the pieces fit together. Civic Progress was no flash in the Mississippi Valley. Banfield and Wilson (1966:267) refer to the development of what may be termed supraorganizations of businessmen after the second world war. In Chicago it was called the Central Area Committee; Pittsburgh: the Allegheny Corporation; Boston: Civic Conference; Philadelphia: Greater Philadelphia Movement. Not mentioned by Banfield and Wilson is the Area Progress Council of Dayton. Before unfolding further arguments, it is important to recapitulate some social science history.

The St. Louis Metropolitan Survey heralded the beginning of a decade of studies and action programs. Since it was an extensive effort on the part of policy scientists to change the political structure, it is worthwhile to examine their evidence for understanding about who made critical decisions regarding the development of the plan to integrate the metropolis and, during the campaign, to convince the electorate. Two local reports (*Background for Action,* Metropolitan Survey, February, 1957; *Path of Progress for Metropolitan St. Louis,* Metropolitan St. Louis Survey, August, 1957) gave the social science facts, and the recommendations for a district plan which would have unified governmental operations in the city and county if it had been adopted. After the voters turned down the district plan, Bollens, (1961:93–94) who had directed the Survey, commenting on the lack of coordination in the city-county area, said that the community lacks the means for developing consensus. It is virtually impossible, as he sees it, to attain agreement through voluntary organizations. He writes,

> This consensus involves two aspects: agreement on what matters are of area concern and agreement on how each is to be handled. In addition, legal and institutional means must exist for making and executing such decisions. The sheer proliferation of governments in a metropolitan area that does not have an area-wide policy-making agency raises a serious question about the possibility of democratic and responsible citizen control of these matters. . . . Lacking an area-wide government, voluntary cooperation together with informal and private means must be used to reach any decision. Such methods are generally insufficient; they frequently lead to no agreement or to agreements that cannot possibly be carried out within the existing governmental framework. They are, moreover, partial and particular rather than complete and general devices. Without a formal mechanism for central decision-making in a metropolitan area, there can be no way of evaluating the segmented forces of interest and pressure groups in the light of the functional aims and well-being of the whole community. Nor can there be any effective way for the citizen to participate in the decision-making process at the metropolitan level.

Greer (1963:21) says the "civic notables" or "chorus of old men" legitimate community endeavors, but based on his observations in St. Louis, they are not otherwise influential. In connection with the St. Louis Metropolitan Survey, he credits them for validating the issue as worthy of study. These men are designated as "integrative symbols" who have a long-term interest in the downtown and the area as a whole, and to whom metropolitan government is a means for protecting and enhancing their holdings (Greer, 1963:24–25). Even though Greer was cognizant of the importance of metropolitan government to utilities such as Union Electric, Laclede Gas, and Southwestern Bell, he does not note that the chief executive officers of these corporations are in Civic Progress, Inc. He acknowledges that these notables, Civic Progress he mentions specifically, raised money too for the Metropolitan District Campaign which was the outgrowth of recommendations of the Metropolitan Survey. This money came from big corporations and he comments that the size of these corporations gives their chief executive officers de facto leadership (Greer, 1963:27). Although giving endorsement and money, Civic Progress did not give of themselves, Greer (1963:62) says:

> Instead, it seems that the notables of the first water shrank from the glory and after several had declined, turned finally to a political unknown. He was a person with a reputation for good works, a war hero who had sponsored the local "boys town." An unexceptional young man, he had no history or particular competence in local political affairs; the campaign was his baptism of fire. Both the abnegation of the "Big Mules" and the choice of this man as leader indicates the contingent commitment of big business to the campaign.

The St. Louis District Plan campaign is used by Greer (1963:62) to test elitist theory. His hypothesis:

> If there were a community power structure in the St. Louis area, it could be expected to either (a) favor the plan and see that it carried, (b) oppose the plan and see that it was defeated, (c) take a "hands-off" position and fail to appear in a chronicle of major events.

Civic Progress did approve and endorse the District Plan. It also formed the City-County Committee to which a number of notables gave their names, and substantial funds. However, the elitist theory breaks down at this point, Greer notes, because they did not exert pressure on the politicians. Moreover, none from Civic Progress accepted the overall chairmanship. A. J. Cervantes became chairman of the City Committee and the County Committee was chaired by a relatively unknown person. Finally, after no notable accepted, the overall chairman selected was the aforementioned automobile dealer. Mayor Tucker's ambivalent role is mentioned by Greer. At the time Civic Progress gave its support, Mayor Tucker had not announced his opposition. During the campaign, he did come out against the proposal. Greer interprets this as also disproving the elitist position because Civic Progress and the Mayor had been very close on many issues over a number of years. Why were they not able to persuade him to go along, Greer asks if they are a ruling group? According to Greer (1963:62–63): "Civic Progress, rather than corralling the mayor, seemed instead to be dependent on him, the wind disappeared from its sails at that point, and new money ceased to flow in. (Indeed, some members of Civic Progress indicated that if they had only *known* Tucker was opposed they would never have favored the District Plan in the first place.)"

Greer's test of the elitist position suggests these questions: (1) What was Civic Progress' reaction to not having any of its members on the Board of Freeholders? Earlier, Greer (1963:25) writes: "In St. Louis, a complex and curious system resulted in nominations by circuit judges (with seconds by the city's mayor and the suburban county's supervisor) and one man picked by the governor." Later, Greer quotes (Schmandt, et al., 1961, quoted in Greer, 1963:45) a description of the make-up of the Board of Freeholders:

> Most of the members were unknown to a wide public audience and few of them were recognized as community leaders. No one of major stature in local government and politics was included among the appointees; neither was the elite of the local business structure. The influential Civic Progress Incorporated, and the top leadership of the Chamber of Commerce were not included. Key officials of civic organizations were also missing. Only the officialdom of the unions was well represented.

In this test of the thesis he apparently overlooks empirical data which he presented earlier in his book that suggested that juridical constraints limited Civic Progress' influence in the appointment of the Board of Freeholders. (2) What would have been the reaction of the community to Civic Progress or to Mayor Tucker if both had opposed the Plan? How much choice did they have? (3) Did Mayor Tucker protect Civic Progress rather than as Greer suggests Civic Progress' being corralled by Tucker? If Civic Progress exerted no pressure on Tucker, could this be interpreted as a sign that it was in agreement with him? (4) Does Greer's formulation really reflect the nuances of power and its relationship to community action? He refers previously (Greer, 1963:29), but does not mention in his test of how influential Civic Progress was, that it only gave a "contingent commitment." Why should they back an issue almost foredoomed to failure? These elites are pragmatists, not reformers.

Greer (1963:200) in concluding, alludes in somewhat somber tones to the metropolitan reforms which have failed:

These studies have underlined the difficulty of bringing about rational change in a rational manner. However, there are many processes at work indicating that change will take place in the future—whether by rational decisions in referenda, by brainwashing and blackout, by administrative fiat, or by shifts in power to higher levels—the government of the state or the federal union itself.

Greer also overlooked the interrelationship of Washington University and St. Louis University with Civic Progress, Inc. Both universities depend for their existence on the noblesse oblige of these civic giants. Their boards of trustees are studded with these elites. Of course, these colleges repay the corporate elites many-fold by serving as a huge brain trust and factory. Greer looked at individuals in Civic Progress per se. He did not mention the interlocking of Civic Progress to the major social policy planning and implementation agencies of the community.

Moreover, Greer did not indicate that the commitment for the metro survey came before the charter defeat in 1957. The St. Louis survey report was released in August of 1957. The failure to achieve the Charter revision could have meant that the passage of the District Plan in 1959 would not have amounted to any appreciable change in area governance anyway. It is not unlikely that Civic Progress drew back their forces for the greater battles that were shaping up for the 1960s. To keep power, it must not be continually squandered in futile campaigns. Finally, the real power, as they demonstrated, rests with the finpolity.

From a functional viewpoint, the findings of policy scientists served the finpols well. It sounded for oncoming generations of social scientists and the lower level of the power structure, that nobody really rules. This was worth every bit of money the Ford Foundation gave and business elites contributed to the abortive campaign.

CONCLUSIONS

The 1950s were a watershed vis-à-vis polity and finpolity. Polities failed to meet city problems and consequently private governments moved in to fill the vacuum. Do not the latter profit from fragmentation of polities? What sense would it make for an interorganization of finpolities to create a large and coherent polity which their own organization might not be able to overpower? Small municipalities in the St. Louis Metropolitan area are easier to deal with when private polities seek extension of domains by land acquisitions. The ninety-five polities in the St. Louis area can vie for favor of finpolity financial lords. Extension of finpolities can add to tax duplicates. Future metro reformers should re-read *The Prince*.

The enigmatic findings of metro reform policy scientists are a function largely of excluding economic and sociological variables. An ideological gap exists among political science, economics, and sociology based on types and weightings of indicators to assess who governs. Political scientists derive concepts and measures of community power relations from their focus on society as a multiplicity of organizations and individuals. Emphasis on concepts which are part of their specialized competence, precludes analysis of institutional interrelationships in a community's power structure. Thus, political scientists, in terms of their commitment to a model, find a reign of pluralism with fragmented power. Economists, although dealing with concentration of economic power, fail to analyze the social structure in which monopolistic devices operate. They do not gauge the interlocking of corporate power with politics, social welfare, civic and religious activities. Hence, they do not take into account the reciprocities and rewards which can be observed by study of social structure or relations among community agencies.

The sociological frame of reference supposes that power is concentrated in the hands of interrelated individuals and agencies. Hence, a conceptual framework focused on social structure is employed to analyze, through a case study, organization and ideology of voluntary welfare agencies in St. Louis. Hopefully, the contribution of this book comes from an attempt to increase understanding of the process in which social policy influence is centralized through the relations among agencies making strategic decisions.

Chapter 4

The Finpolity at Bay

Civic Progress, representing large finpolities, was hard pressed by Washington Federals and black power during the tumultuous sixties. The validity of private ideology was challenged but in spite of ghetto conditions which seem to stay essentially the same, the image of Civic Heroes remained pretty much untarnished at the close of the decade. The Federals, symbolized by LBJ, appeared at times as pathetic Don Quixotes, and toward the end came on bended knees asking for succor from the wealth of finpolity kings. Noblesse oblige of the finpolity welfare system was assured, and the public polity will use whatever means necessary to crush the politics of antireason, whether expressed by black, poor, student, or other.

Aside from abortive attempts at metropolitan reform, or political gerrymandering, reorganization of the Community Chest into the United Fund, and activities concerning an earnings tax, bond issues, urban renewal and public housing, the civic influentials—and Civic Progress in particular—could look back upon the fifties as a period of relative calm before the urban storm. These were Eisenhower years, which he himself observed, allowed for great growth of the, now clichéd, military-industrial complex. What Truman started, Eisenhower extrapolated upon and left, as the heritage for Kennedy, the rudiments of the Vietnam War. The "Redbaiting," the alleged Communist menace, the nuclear stockpiling, and the global militarization by the United States had reached about as far as it could go without starting World War III. In spite of having American eyes looking outward to supposed dangers from abroad, and exacerbating the international situation in the early sixties, the domestic scene was becoming decidedly troubled.

Martin Luther King, Jr., and the students, had been active during the waning years of the fifties, but it was not until 1960 and 1961 that the decisive break came which led to the Negro uprisings (National Advisory Commission on Civil Disorder, 1968:227). The transition came with student sit-ins which ended in arrests and failure.

THE NEW FRONTIER

It began as the youth decade led by a great phrase-maker of the century. Kennedy, the idealist-realist, like his predecessors, kept his rendez-vous with military power. The gap between his ideals and the politics of realism, both foreign and domestic, was glaringly apparent.

Black militancy fed upon such diverse events as voter registration efforts of 1961; the breaking off of the Student Non-Violent Coordinating Committee from the Southern Christian Leadership Conference and the National Association for the Advancement of Colored People, and James Farmer becoming head of the Committee on Racial Equality. Even Whitney Young, Executive Director of the National Urban League, took a stiffer line with business elites who heretofore had been handled with esteemed propriety.

This was also the year of the Juvenile Delinquency and Youth Offenses Control Act. It provided funds for communities to develop "techniques to prevent and control juvenile delinquency to promote youth development" (Office of Economic Opportunity, 1967:310). In addition, it called for coordination of public and private programs concerning delinquency. With the passage of this Act, package programs became the distinctive feature of private and public welfare systems in the sixties. The Act is even more important because it, along with the Ford Foundation's Grey Area Projects, brought a censorious consciousness of inadequacies in private and public programs. Foundations and the Federal government scrutinized Establishment agencies.

The Ford Foundation originated the Grey Area Projects as a challenging substitute for metropolitan reform (Marris and Rein, 1967:15–16). This quixotic social science crusade had just about ended by 1959. The school system seemed to be a fulcrum for lifting up those listless, grey-looking areas of the cities lying between the glistening downtowns (thanks to urban renewal projects) and the affluent white suburbs. The schools could be opportunity and adaptation centers for children of Appalachian and black migrants.

Schools could, too, become intimately involved with the community they served—educating not only the children, but their parents, and, by a program of evening activities, the adult population at large. So the Foundation saw in the school system of the city an opening for its policies which it eagerly exploited. In March 1960, an initial grant was proposed of one and a quarter million dollars to seven school systems—Chicago, Cleveland, Detroit, Milwaukee, Philadelphia, Pittsburgh and St. Louis (Marris and Rein, 1967:16).

Schools and social agencies were criticized for both ineffective socialization and failure to modify opportunity structures of those called the culturally deprived. Established organizations, proponents of the new programs asserted, did not comprehend the interrelationship of human problems and focused excessively on individuals rather than on home, peer group and neighborhood milieus. Also coordination of available community resources was said to be lacking (Kravitz, 1968:265).

The surging civil rights movement was concomitant, of course, with both school and delinquency projects. All three were problem, rather than agency, oriented. The problem focus became another cliché of the sixties. The multiproblem family was an ardent interest of some social workers. Demonstration Projects in Public Assistance under the Public Welfare Amendments of 1962 were instituted (Office of Economic Opportunity, 1967:272). Also in this same year the Manpower Development and Training Act was passed.

Public housing projects built during the fifties in St. Louis were publicly recognized as failures about this time. The high-rise mausoleums, unsuitable for families with small children, were also ghettos for the pooling of one-parent welfare families; the physical design invited crime and vandalism. In 1962, Alvin Gouldner, a sociologist at Washington University, obtained a National Institute of Mental Health Grant to study the social life of a public housing project. The subject of this $300,000 study was high-rise Pruitt-Igoe with its more than 2,000 families (nearly 11,000 individuals) having an average annual income of $2,500 and nearly fifty percent receiving public assistance (Gouldner, et al., 1966). This event is important because it was another thorn in St. Louis' hide, a further source of embarrassment to civic elites. Although this could be absorbed by public officials, as customary unpleasantnesses usually are, discontent, high crime rates, health and school problems reflected unfavorably on St. Louis.

Pruitt-Igoe was to be a community laboratory for comprehensive services. Officials promised that the juvenile delinquency control, the across-the-street city health center functions, and the St. Louis Division of Welfare would move ahead together in relating to people's problems. In addition to the Washington University study, the Division of Welfare's Pruitt-Igoe

Demonstration Project sponsored a study of ADC (Aid to Dependent Children) Families in 1962 (Division of Welfare, 1964). This Committee was tied to Civic Progress elites through Robert E. Hillard, of Fleishman-Hillard, Inc. (firm used by Civic Progress for public relations needs), and Ethan A. H. Shepley, Jr., Vice President, Boatmen's National Bank (this bank was represented in Civic Progress by Harry F. Harrington).

Civic Progress did not stand still during 1961 and 1962. It had taken over hospital planning; campaigned successfully for the $95 million Metropolitan Sewer Bond issue; also backed other bond issues. Besides these achievements, Civic Progress worked for tax increases for the public library, zoo, and art museum. Through Downtown Inc., Civic Progress spearheaded Mansion House (a block-long high-rise luxury complex of apartments and businesses on the riverfront).

United Fund service and planning agencies, youth commissions, public welfare agencies, health departments, and housing authorities were often by-passed by the federal delinquency planners. It was the spirit of the young rebelling against the bureaucracies which fifty years of social work had created. Brokers of civic influentials, keepers of the poor, and especially welfare, recreational and health dispensaries for middle classes were publicly denounced as not having the credentials for leading disadvantaged youth of America across the New Frontier.

The hypothesized influence structure was not unchallenged. It was opposed, but was it really endangered? Was there an ascendancy of public ideology genuinely reflective of the public interest? The following were indicative of opposition to the Establishment, and also of the latter's effort to defend itself.

The Rockefeller Foundation in 1958 appointed an ad hoc committee to study the place of private agencies in the United States. It was concerned over the multiplicity of agencies, competition among fund raising groups, and the participation of the federal government in health and welfare ventures. The questions studied (Hamlin, 1961:4) were:

> Are voluntary agencies necessary, in view of the great expansion of government health and welfare activities? Can the relative responsibilities of government and voluntary agencies be defined? What should be the relationship between government and voluntary agencies? Should government supervise voluntary agencies in the public interest by requiring full disclosure of agency income and expenditures?

The federal tax proposal of 1962 would have allowed deductions for contributions, union dues, sales taxes, state income taxes and the like only if such items exceeded five percent of one's gross income. Agencies soliciting money for charitable purposes viewed this as a major threat to private philanthropy.

The issue ran deeper than a matter of gain or loss of revenue, but instead went to the heart of public policy, which has historically encouraged very strongly the growth of private philanthropy and the consequent development of a pluralistic and complementary system of public and voluntary social welfare (Manser, 1965:828).

St. Louis was in step with other vanguard cities, and like New Haven, sought to be in the forefront of the new federal programs. The Juvenile Delinquency Control Project, initiated by Mayor Tucker with Civic Progress endorsement, had the technical assistance of Professor Nicholas Demerath, sociologist from Washington University, as director of the project. Although the initial phases were coordinated with the St. Louis Division of Welfare, Health and Welfare Council (arm of the United Fund) and the Metropolitan Youth Commission, then under the Mayor's Office, this project was organized as a separate entity following the philosophy of David Hackett of Bob Kennedy's staff.

This was 1963 and violent times had begun. There was trouble in Birmingham—civil rights workers were killed. Chicago and Philadelphia had disorders. It was the year of the March on Washington for civil rights. CORE was getting blacker, and the battle hymn for black Americans was "freedom now." It peaked with the assassination of John F. Kennedy.

Civic Progress influentials further tightened their control over social policy. The United Fund (UF), dominated by Civic Progress, cracked down on the Health and Welfare Council (HWC) for not attending sufficiently to the business of United Fund agencies, being too involved with federal projects, and recommending large expenditures of money for community services, based upon HWC's Resource-Needs Project Study, started in 1961 and completed in 1963. They ordered HWC to review all the programs of UF agencies, and took budgetary control out of the hands of the executive director of the Council. Most of the Council's budget was earmarked for agency reviews.

Also, Civic Progress set up the Regional Industrial Development Commission to attract more industry to the St. Louis area. This new smaller agency could break away from entrenched interests of the Chamber of Commerce which protected St. Louis establishments not infrequently by resisting competition.

Militant St. Louis blacks demanded white collar jobs from the Jefferson Bank and Trust Co. that year. This controversy, starting in August and lasting until March of 1964, was inspired by the St. Louis Committee on Racial Equality.

Briefly this is what happened: On August 30, CORE picketed the bank alleging unfair hiring practices; some people from the picket line entered the bank, several locked arms and blocked the door, others sat down singing and clapping. Bank officials expecting something of the kind, had been to court and obtained a temporary restraining order to block any interference with their business. Nineteen people—nine as a result of that incident, ten more after later demonstrations—were tried, found guilty of violating the court order and sentenced to various fines and long terms in jail. Picketing and demonstrations continued, and others were arrested on lesser charges (Brodine, 1964:12).

This represented a severe rejection of this type of protest as a strategy for job rights. Such behavior was judged criminal, and its leadership discredited:

When the political life of Negroes was circumscribed by the NAACP, it was clear that political life entailed normative behavior with the formal civic culture. Similarly, it was clear that acts of personal deviance fell outside the realm of politics. Indeed, there was little contact between Negro deviants and participants in the civil rights protest. The rise of civil disobedience entails personal deviance to attain political ends. Regardless of the political goals involved, it is conscious violation of the law (Horowitz and Liebowitz, 1968:286).

Thus the crucible for converting political deviance into the social welfare model was readied.

Although community leadership was moved to open up white collar bank jobs, St. Louis whites were generally numbed by the events. The black community united behind the convicted. Thirty-five Negro leaders declared:

Either the Negro citizens of this city are to continue as a peripheral insignificant segment, limited or excluded by practices of racial segregation and discrimination, or they are to be included as an integral part of all aspects of community life. The latter decision must be the choice of our educational, political, industrial and civic leadership, or our city will face continuing strife and decay (Brodine, 1964:24).

THE GREAT SOCIETY

The Other America (Harrington) was published in 1963, giving poverty a momentous thrust into public awareness, and by the time the Economic Opportunity Act was passed in August 1964, the new St. Louis Human Development Corporation (HDC) was ready to coordinate the War on Poverty in St. Louis and St. Louis County. HDC, so named in anticipation of what was to come, evolved from the Juvenile Delinquency Control Project. All federal programs relating to health and welfare were to be cleared through its offices. Plans were to be locally prescribed and the main support would be federal. However, it was emphasized that the corporation would be independent and free from partisan politics (Memorandum from the Board of Directors, St. Louis Human Development Corp., September 24, 1964).

The first board of HDC had J. Wesley McAfee of Union Electric (in Civic Progress) and three young business influentials: Raymond Wittcoff, President of Transurban Investment Corporation; Ethan A. H. Shepley, Jr., Vice President of Boatmen's National Bank (his father was former chancellor of Washington University, unsuccessful candidate for governor of Missouri, and emeritus in Civic Progress) and Edward J. Walsh, Jr., investment broker; Judge Theodore McMillian (circuit court and a black) and ex officio members: Governor of Missouri; Mayor of the City of St. Louis; Supervisor of St. Louis County, and James Doarn, Director of Region VI of HEW. In addition, there were the appointive political officials: William Kottmeyer, Superintendent of St. Louis Public Schools, and H. Sam Priest, President of Board of Police Commissioners of the St. Louis Police Department (the Board of Police Commissioners for St. Louis is appointed by the Governor as prescribed by statute). Finally, there was a representative from labor: Joseph P. Clark, President of the St. Louis Labor Council, AFL-CIO; and Mrs. Irvin Bettman, Jr., wife of a prominent businessman. The general manager was Wayne Vasey, Dean of Washington University's School of Social Work on leave of absence.

Professor David J. Pittman was consulted by the author (St. Louis, Mo., November 20, 1967) and indicated that Professor N. J. Demerath of Washington University had assembled the list of names for HDC's first board for Mayor Tucker who went over the names with Civic Progress members. This is indicative of interlocking technocrat, political and business roles.

The foregoing materials indicate that (a) The Human Development Corporation and its precursor, the Juvenile Delinquency Control Project were originated from federal legislation with the imprimatur of the Executive Office of the Presidency; (b) The Juvenile Delinquency Control Project (sponsored by HWC, UF, the Mayor's Office, the Metropolitan Youth Council, and Washington University) constitutes an interrelationship of private and public influentials at national and local levels counseled by experts in the policy sciences; (c) The main stimulus for the social welfare model was the Negro Revolution and fear of the consequences which put civil rights as issue number one in the United States and in the big cities; (d) The interests represented nationally, and in St. Louis in HDC, were not dedicated to changing the social network of voluntary agencies, but rather to making adjustments without altering the sociopolitical structure.

The power elite agreed to and participated in the formation of the Human Development Corporation to take advantage of the Economic Opportunity Act. This new agency was set up separately from the city government as a nonprofit, voluntary agency. It was quasi-public because of dependence on federal funds and having its operation defined by the Act itself and by administrative interpretations of the Act. In addition, the Mayor and Supervisor of St. Louis County were given the right to appoint a specified number of members to the board of directors; others were to be elected by representatives of the poor. Thus a structural basis developed making for disharmony between and among nonlocal and local cadre and elites. Nonetheless, civic influentials had board control directly and indirectly. Both Mayor and Supervisor need these St. Louis gentlemen.

Of course, corporate influentials in St. Louis were displeased with HDC because it cast federal officials in the role of telling them and the established agencies what they should be doing. The cadre of the new agency, because of receiving virtually all of their salaries from the federal government, were in a position that was less easy for corporate elites to dominate. On the other hand, the matching money had to come from the local community, and this is the power of moneyed interests.

The foregoing is borne out by Civic Progress' holding back money from the newly formed enterprise. Their position was one of "let's-wait-and-see." Business elites did not want to increase the "spending rates" of agencies with "soft" money.

People were asking questions: "Does this mean an end to some UF agencies?" "Will the OEO programs hurt the UF campaign?" "How effective are the community-supported agencies anyway?" The protective veil was ripped off elitist charities and philanthropic undertakings; public agencies were also scored. But the people doing this carping (HDC staff, social scientists, some blacks, and lower echelon social workers) were not particularly influential in St. Louis power circles.

Populist, public ideological, currents crested on the Mississippi. "Maximum feasible participation of the poor," written into the Economic Opportunity Act, chilled the blue blood of elitist St. Louis. Sixty-four was also the year of the Civil Rights Act containing such populism as public accommodations and equal employment opportunities. Johnson was elected President and the Great Society was upon us. Social scientists and social workers of national renown, bolstered by Johnson's boldness, produced a barrage of negative criticism of Establishment community and public welfare systems.

According to S. M. Miller (1964:302–305), both public and private social agencies had not met needs or been sufficiently accountable to the public. He pointed to eight basic shortcomings: (1) They have been generally inadequate, with an overreliance on private agencies; (2) Coordination has been lacking, and agencies often inaccessible to clients; (3) Middle class clients have been preferred with a consequent underemphasis on services to the poor; (4) Priority has been given to custodial and remedial rather than preventive type services; (5) Individualized service has been limited because of the emphasis on moral and fiscal matters; (6) Staff often have failed to meet needs of the poor; the highly trained are psychotherapeutically oriented, and the untrained little more than policemen or bookkeepers; (7) Solicitation of funds has become a substitute for public action and "private government" has developed which is not sufficiently accountable to the public. Agencies have been oligarchical and unrepresentative of the poor. The "tax base" is regressive. The poor have been excluded from policy decisions on what services would be funded or how those funds should be distributed. Agencies have become colonial administrators to the natives; and (8) The poor have not been organized to help themselves; instead the emphasis has been on their using community services.

The shortcomings in private planning have been summed up differently but in the same spirit by Solendar (1963:3–24). There is a lack of comprehensive planning for health, welfare and recreation; poor relationships with public agencies, and planning an agent of the United Fund from which its major support has come; unrepresentativeness of boards; serving middle and upper classes better than the poor; and the need to relate planning to the federal resources.

Schottland (1963:97–120) reviewed briefly the history of federal planning, and maintained that the role of government will continue to grow in size and influence. Local social work leaders were chided for being uninformed about the importance of federal planning. Schottland was unafraid for social planning done by local health and welfare councils, because he saw the problem not as competition but learning to work together to further the mutual interest of both. The federal government, Cohen (1964:3–19) states, should give more leadership to the private agencies. According to Cohen, the Welfare State is only a symbol which takes our eyes off the problems; debates over public vs. private are fruitless, and the need for comprehensive centralized planning at all levels is essential.

Defensiveness on the part of the cadre of the corporate influentials is suggested in the following citation. Edelston (1964) has noted confusion regarding roles and loss of traditional initiative as federal planners and federal funds increasingly come into local communities. "No longer," he remarked, "are communities rulers of their destinies." C. F. McNeil (1964 and 1965) distinguished between administrative planning and community planning. The latter is the proper function of the local community and involves coordination of lay and professional groups relating to social welfare needs, problems, and resources. Having the government plan "all human services," said McNeil, "is different from government's planning government's services."

United Community Funds and Councils of America (an association of United Funds and Welfare Councils) launched a nationwide study of private agencies because of the challenge presented to these agencies by the Office of Economic Opportunity. The report (Guidelines for Local Participation UCFCA Voluntarism Study, 1965) stated that

> the establishment of organized health, welfare, and recreation service under both voluntary and governmental auspices is being challenged as never before. In some cases, existing welfare activity is being by-passed or ignored. There are those who say that much of it is irrelevant. There is no question that some institutional paralysis exists and that imagination, innovation, and in some cases major surgery are needed as never before.

The report concludes that "these are certainly evolutionary and probably revolutionary times for social welfare. Most revolutions result in the destruction of established institutions, at least those that are rigidly resistive to change."

A major problem of the "nongovernmental" organization, according to Alan Pifer (1966), acting president of the Carnegie Corporation, is how to get a retainer from the federal government and still be independent.[3] He did not think it insoluble and called for a national study of the role of the private agency which he believes is essential to government. Complexity of national and international problems requires resources which government could not build into its bureaucracy, he asserted, and therefore, service must be purchased. Misuse of private agencies by the government is attributed, primarily, to the federal government's not providing these agencies with a steady source of income. There is the further problem of being free to say "no," but being afraid of the consequences if one does, Pifer comments.

> Is the nongovernmental organization of the future to be simply an auxiliary to the state, a kind of willing but not very resourceful handmaiden? Or is it to be a strong, independent adjunct that provides government with a type of capability it cannot provide for itself? (Pifer, 1966:14).

Private funds are not sufficient to finance these agencies according to Pifer, and he envisions a new mechanism—a federal united fund. Pifer would have the organization be not a "whore" but a "wife."

BLACK REVOLUTION AND THE GREAT SOCIETY

Nineteen sixty-five was a revolutionary year. The Watts riot and the expanding black militancy were ubiquitous. The Southern Christian Leadership Conference marched on Selma for voting rights and there was the March on Washington for Peace.

In addition to the pronouncement of the Great Society, the federal legislation was legion: Housing and Urban Development Act; Civil Rights Act on voting; Higher Education Act; Elementary and Secondary Education Act; Public Works and Economic Development Act, and Medicare. The first *Catalog of Federal Assistance Programs* was published to spread the Great Society's glad tidings.

The War on Poverty both nationally and locally was continually being challenged for wastefulness, high salaries, corruption and fomenting disorders. The *St. Louis Globe-Democrat* crusaded against it and started its drive for the Herbert Hoover Boys' Club.

St. Louis had a new mayor and he was elected, in part, through efforts of black militants. Alphonso J. Cervantes, self-made man from taxis and insurance, had defeated Raymond F. Tucker, "Mr. Clean" compared to Cervantes and long-time favorite of Civic Progress. Although the latter did not quite trust Mayor Cervantes because of his connections with underclasses, he superseded Tucker as an ex officio member of Civic Progress.

That year Civic Progress announced civil rights as the major community problem. Through their agents in the United Fund and the Health and Welfare Council, Civic Progress further consolidated their holdings by putting the Hospital Planning Commission and the HWC under the same executive director. Later that year the Metropolitan Youth Commission was also placed under the executive director of HWC and HPC.

The following year, 1966, was another page in the Great Society: Demonstration Cities and Metropolitan Development Act; Adult Education Act; Comprehensive Health Planning and Public Health Services, and Child Nutrition Act. But, this year saw riots in Omaha, Atlanta, Dayton, and many other points. Meredith marched from Memphis to Jackson; black power became another term in the sixties lexicon and King announced SCLC would begin its northern drive.

St. Louis with the help of Civic Progress got the general obligation bonds passed. Employees of the Human Development Corp. worked in this campaign. This was also a year of increasing emphasis on employment of blacks and Civic Progress set up Work Opportunities Unlimited with a modest amount of their own funds matched with OEO money. They did this on their own terms with the purpose of showing the effectiveness of a businessman's approach. This, of course, adumbrated what was to come in 1968 with the National Alliance of Businessmen.

In spite of a bus boycott and a threatened rent strike at Pruitt-Igoe in 1966, St. Louis had no riots. The boycott and strike were eventually put down. The alliance of key black and white politicians with chief executive officers of Civic Progress was everywhere apparent. The Human Development Corporation served the community leadership well by giving jobs to blacks who had been leaders in demonstrations. The above-mentioned instances suggest that federal legislation of the 1960s widened the scope of social policy and substantially increased the vertical dimensions and tensions of interagency relations. Community agencies have many more interactions and conflicts with state and federal agencies related to implementation and coordination of new programs. Federally sponsored voluntary agencies (de jure voluntary but de facto public) complicated the already intricate web of horizontal interagency relations at the city level. Agency anarchy appeared more pronounced than heretofore. Relations between public and voluntary agencies were for the most part not rationalized by juridical and administrative stipulations. The pattern of relations was established either through persuasion, or coercion initiated by superordinate voluntary, public or business organizations. Consequently, tensions, conflict, uncertainty, and unpredictability were manifold in urban planning.

Hauser (1967:6) glosses over the confusion and disagreements in proclaiming the birth of the American Welfare State:

> The very fact that this paper on social goals has been requested by the Health Forum is, in itself, not only an indication of the changed character of our society but, also, an additional piece of rapidly growing evidence that the time has come for the formulation of comprehensive policy in respect of social goals and a coordinated and integrated series of programs to achieve these goals. In stating this position it must be recognized that, in effect, it is being assumed that the United States is no longer an agrarian society characterized by a laissez-faire economic outlook and the conception that "that government is best which governs least." On the contrary, it is being assumed that the United States as an urbanized and metropolitanized society has come to understand that the personal, social, economic and political freedoms enjoyed by its populace can and must be enhanced by positive government interventionism as necessary for the welfare of the American people. In brief, it is assumed here that the United States is a welfare state and that such a designation is neither pejorative nor

dangerous. It is rather a badge of maturity—explicit recognition of the changed character of American society and the new requirements by reason of the change.

Hauser's position is like many policy scientists who, troubled by the urban scene, advocate a centralized approach.

In spite of Hauser's optimism, planning under governmental auspices often means that solutions to problems fail to give those to be helped a sufficient voice, and programs are niggardly. Even though programs stipulate representation of the "poor," this may be specious in terms of who really continues to decide policy at the local level. This can result in a worse situation than before where social policy instead of societally evolving along nonrational lines is juridically regulated with democratic facades, making fundamental changes more difficult. Although rebellion may be quelled, the basic underlying problems are still unmet, and thus legislating representation for the poor becomes another form of paternalistic oppression. Finally, the tax supported health and welfare services are inadequate. Income provisions and social services are based on minimum standards. To be otherwise would require substantial increases in taxes.

Although the Economic Opportunity Act has shades of public ideology in its provisions for maximum feasible participation in determining how poverty money is to be spent, this was only implemented in a token sense. Thus, elitist control of decision making was not significantly lessened.

The heart of the ideological conflict and tensions introduced by the federal programs was who would rule. The issue was not whether there would be federal aid as far as dominant interests of the St. Louis power elite were concerned. That was more a bother to conservatives at the national than local level.

The periods of 1967 and 1968 were noteworthy in St. Louis' continuing to be riot free. This, of course, was not the case with many other cities, even across the river in East St. Louis. The Riot Commission was appointed and the summer was fiercely contended in black ghettos of Detroit and Cincinnati.

The cooperation between the Democratic Mayor and Civic Progress continued. They backed him on the Spanish Pavilion and bringing the reconstruction of the Santa Maria to St. Louis. The business and political leaders moved ahead to apply for a Model City grant.

In 1968 President Johnson instituted the National Alliance of Businessmen to deal with black unemployment. It was apparent by that time that OEO and the Department of Labor, which had been continually dueling over jurisdictional matters both nationally and in St. Louis, would not be able to deal satisfactorily with a generally deteriorating situation.

St. Louis had a model in Work Opportunities Unlimited. The businessmen, who for a time had been at bay, saw in the decline of HDC and public recognition not only locally but nationally, that Big Business in America is socially responsible. Moreover, the principle of corporate or finpolity welfare was vindicated and governmentally packaged programs were starting to float down the Mississippi.

King and the second Kennedy were assassinated and America had a Republican president again.

CONCLUSIONS

The tale of the sixties is the black revolution for which Civic Progress sought to gird its political loins in the fifties. The national corporate influentials (as some Civic Progress members are) assumed command of reform forces. With their charter from the federal government, a national conglomerate of finpolities organized to fight for the basic freedom, as defined by private ideology: that is, the freedom to seek invidious pecuniary gains. This union of finpolity and polity for purposes of maintaining a social and economic system dedicated to the principle that social policies of inequality are necessary represented a unique development in America's evolution toward totalitarianism of the center.

The Corporation-State conglomerate is at bay. The myths of the sixties such as the War on Poverty, the Model City Agency, black capitalism, and the latest, JOBS of the National Alliance of Businessmen—also in the context of the welfare model—are myths because they promise equality and deliver inequality. Some students and blacks realize it cannot be otherwise when the basic and real commitment is for inequality in order that there be incentives to sustain and increase productivity.

Will this conglomerate allow the basic freedom to be substantially subverted? So as to secure this basic freedom against assault, what will happen, in the Morgenthau-like and Niebuhr-like politics of realism, to secondary freedoms—speech, press, assembly, and their various forms as embodied in protest?

Part Three

CORPORATE

AND COMMUNITY

WELFARE

Chapter 5

Civic Progress, Inc.

Up to this point I have dealt, in a general way, with the nature and tensions of policy and agency. The central agency in this case study of St. Louis is Civic Progress, and its policy is enlightened private ideology. It had in 1966 only thirty-one members: ten manufacturers; six bankers; three retail heads; three utility chiefs; two university presidents, and others scattered through key positions in the finpolitan complex. To those who mistake shadows in the cave for real substance, such an agency and its members are not private men, who serve first the interests of their corporations, but rather public spirited citizens positioned above the throng often because of hard work and "Yankee know-how." These men are cherished and lauded because without them the city could be worse. In the fifties Civic Progress was founded, as previously mentioned, because Mayor Darst was having difficulty getting anything done. The sixties tested its mettle and the 630 foot stainless steel arch completed in this decade is perhaps as much a tribute to civic influentials as it is a symbolic gateway to the West. A democratic city polity was found wanting, not Civic Progress.

What lies ahead in this chapter is an attempt to understand facets of Civic Progress' power. This agency, of singular significance to St. Louis, is a prototype for other American cities. It merits a much more detailed analysis than the facts at hand permit. If my assessment is valid, organizations like Civic Progress should be continually monitored by underclass agencies which do not have such direct dependence on the noblesse oblige of these civic noblemen as do the Establishment agencies. To be above trenchant and consequential criticism is to hold an exalted position not unlike the medieval lords and clergymen. How this could happen in America, the presumably classless society, is paradoxical. By swamping with factual criticism from various sources, perhaps, we can recapture the public ground which slipped by default into the "new estate."

The power source of Civic Progress lies partially in the corporations which are its members. This power is simply the potential to use force to obtain welfare objectives. Civic Progress is extraordinarily powerful because of the ability of its corporations to use force. This gives it a uniqueness as a private community welfare agency.

Ordinarily, the State is considered to have a monopoly on legitimate violence. It is not usually noted that private governments or finpolities also rule by legitimate violence, and this rule can be far more important in every day life than potential effects from public governments. These finpolities maintain the security of their premises; they have their own police forces, private detectives, and most tellingly, they have the power of bestowing or destroying one's employeeship or tenure. Compared to the public polity, it is harder to become a member and easier to be ejected. Hiring may be discriminatory; firing preemptory.

Unions offset finpolity power to a degree, but in return for security afforded, there is a caste status fixing most, more or less involuntarily, to a given range of positions within a set classification—white collar, blue collar, clerical, professional and the like. Moreover, a social policy of unequal welfare: wages and other benefits, is established. This is backed by force of law, and is underwritten by the religious thinking embodied in the Protestant Ethic. What has happened is a fusing of religious and secular in the "calling." The profit making aspect is interwoven with liberal democracy to complete the secular design of private ideology. The union caste system is ruled by an oligarchy of high priests who serve as brokers between corporate nobles or top management and the vassalage.

The power of Civic Progress, as already indicated, is in the corporations but even more important this power is a function of corporate interrelationships and interpenetration (by chief executive officers and corporate underlings) of voluntary and public agencies. That is the crux of this chapter.

FINPOLITIES

Table 1 shows the corporations in Civic Progress. The diversity is seen in the firms included: industrial, merchandising, banking, utilities, insurance, transportation, investments, real estate, engineering, law, and even St. Louis University and Washington University. Several of the corporations are among the largest in the United States in assets and numbers of employees. Most significantly, however, these are among the largest locally based corporations. This is significant from the standpoint of having more of an economic stake in the local domains and of course facilitating opportunities for reciprocities because men at the top can make decisions faster. Nonlocally based corporations do not have their highest ranking nobility in foreign kingdoms.

To grasp better the magnitude of these establishments, Table 2 shows the net sales and national ranks of St. Louis based industrial corporations listed in *Fortune* magazine's 500 largest corporations (*The Fortune Directory,* 1963, 1964, 1965, 1966, 1967). The table presents data only on those eleven companies which have been included in this elite grouping in each of the past five years. These companies in 1966 employed nearly 258,000 and had sales in excess of $6 billion (*The Fortune Directory,* 1967).[4] In *Fortune's* list of the fifty largest banking firms, Mercantile Trust ranked 41st; General American Life Insurance Company was 49th among the fifty largest insurance companies. Table 3 includes banks in this area with total resources exceeding $100 million:

Of the thirty-nine principal banks in St. Louis, the four largest have more than 60 percent of the resources. The frequency table below shows the concentration of banking wealth in relatively few banks (tabulation made from *Manual of St. Louis Bank Stocks,* G. H. Walker & Co., March, 1966):

Resources in Millions	Number of Banks
$200+	4
150–199	1
100–149	2
50– 99	6
Under 50	26
Total	39

Another indication of how a few firms have most of the employees is seen in Table 4. Those employing 1,000 or more, although being less than 10 percent of all firms, have more than 50 percent of the area employees.

"FINOPOLY"

Lines of influence are those interconnections among the corporations and the relationships of these corporations to agencies which make and carry out social policy.

The corporate wealth of St. Louis is joined in Civic Progress, Inc. The

Table 1

CIVIC PROGRESS CORPORATIONS, 1966

	Millions of Dollars	Number of Employees
Industrial	(Sales)	
Monsanto	1612	30,000
Ralston Purina	1154	20,741
Anheuser-Busch	479	9,340
Interco	469	32,600
Pet	423	15,400
Emerson Electric	348	17,871
Brown Shoe	301	21,730
Falstaff	168	4,540
Granite City Steel	159	5,597
General Steel Industries	*	4,000+
Banking	(Resources)	
Mercantile Trust	1123	1,413
First National Bank	913	1,093
General Bancshares Corp.	437	1,005
Boatmen's	315	500+
St. Louis Union Trust	30	369
Merchandising	(Sales)	
May Department Stores	984	45,000
Edison Bros. Stores	175	8,400
Stix, Baer & Fuller	*	2,000+
Utilities	(Operating Revenue)	
Southwestern Bell	1198	59,609
Union Electric	209	6,286
Laclede Gas	85	2,345

money and manpower which are the sources of influence of the members of Civic Progress, whether they act individually or collectively, resides in the large "indigenous" or locally based corporations. Civic Progress has among its members the chief executive officers of nine of eleven nationally ranked industrial corporations of St. Louis: Monsanto, Ralston Purina, Anheuser-Busch, Interco, Pet Company, Brown Shoe, Emerson Electric, Granite City Steel, and the Falstaff Brewery. In St. Louis, only one bank and one insurance company made Fortune's illustrious listings for 1966: both are represented in Civic Progress (Mercantile Trust Co. and the General American Life Insurance Co.). Considering the four largest banks

Table 1 (continued)

	Millions of Dollars	Number of Employees
Life Insurance	(Assets)	
General American Life	402	500+
Transportation	(Operating Revenue)	
St. Louis-San Fran. R.R.	152	8,419
Real Estate		
Clarence M. Turley, Inc.	*	200+
Engineer-Architect		
Sverdrup-Parcel	*	500+
Investment Banker		
Reinholdt & Gardner	*	100+
Public Accountant		
Price-Waterhouse & Co.	*	*
Corporation Law		
Judge James M. Douglas	*	*
Education		
St. Louis University	*	*
Washington University	*	*

* Not available.

The Fortune Directory, Fortune, June 15, 1967. Moody's Bank & Financial Manual, April, 1967. Moody's Industrial Manual, June, 1967. Moody's Public Utility Manual, August, 1967.

in the St. Louis area, all of their chief executive officers are in Civic Progress. From the field of merchandising, the May Company is in Civic Progress as are the smaller Edison Brothers and Stix, Baer and Fuller. These three are the largest in this commercial endeavor in St. Louis. The three major utility companies: Southwestern Bell, Union Electric, and Laclede Gas, are in Civic Progress. The Civic Progress companies are also among the largest employers. Table 5 indicates about 83 percent of these companies employ 500 or more people. A list of Civic Progress members in 1966 is given in Table 6.

Table 7 lists sixteen Civic Progress members (one is emeritus) and their intercorporate memberships. The emeritus member, Edwin Clark, formerly head of the Southwestern Bell Company, is included because the

Table 2

NET SALES OF LARGEST ST. LOUIS BASED INDUSTRIAL FIRMS AND THEIR NATIONAL RANKINGS, 1962–1966

	Sales in Millions of Dollars					National Rank				
	1962	1963	1964	1965	1966	1962	1963	1964	1965	1966
Monsanto	$1063	$1193	$1359	$1468	$1612	43	39	33	34	37
Ralston Purina	682	808	864	954	1154	70	65	66	68	59
McDonnell	391	565	865	1008	1060	134	101	65	57	66
Anheuser-Busch	327	344	376	422	479	156	163	163	165	161
Brown Shoe	323	317	247	265	301	161	176	251	258	261
Interco	303	296	346	392	469	171	188	174	173	166
Pet	234	261	291	347	423	222	215	212	196	183
Emerson Electric	217	208	219	253	348	239	260	275	271	229
Granite City Steel	136	140	161	166	159	342	349	343	360	412
Peabody Coal	133	160	189	208	233	348	319	303	310	317
Falstaff	125	132	139	151	168	364	364	375	287	393

The Fortune Directory, Fortune, July, 1963, 1964, 1965, 1966, and June 15, 1967.

new president of this company was not yet listed in *Standard and Poor's*. The First National Bank of St. Louis had eight board members from Civic Progress; St. Louis Trust, seven; General American Life Insurance Company, six; and Anheuser-Busch, five. The most active members of other

Table 3

ST. LOUIS BANKING COMPANIES BY TOTAL
RESOURCES EXCEEDING $100 MILLION

Bank	Millions of Dollars
Mercantile Trust Co.	$1103
First National Bank of St. Louis	851
Bancshares Corp.	411
Boatmen's National Bank	285
St. Louis County National	154
Tower Grove Bank and Trust	147
National Stock Yards National Bank	118

Manual of St. Louis Bank Stocks, G. H. Walker & Co., March, 1966.

executive's corporations were Calhoun who heads St. Louis Union Trust, with eight; McAfee of Union Electric with seven, and General Sverdrup of Sverdrup & Parcel who had five board memberships. Based on the fore-

Table 4

NUMBER OF FIRMS BY EMPLOYEE SIZE, ST. LOUIS
METROPOLITAN AREA, 1966

Size of Firms	Number of Firms	Approximate Number of Employees
Over 4,000	14	109,908
2,000–3,999	17	46,102
1,000–1,999	39	53,678
500– 999	82	55,361
400– 499	66	28,505
300– 399	81	27,460
200– 299	152	35,566
100– 199	283	38,493
Totals	734	395,073

Metropolitan St. Louis Large Employers, 1966, Chamber of Commerce of Metropolitan St. Louis.

going facts, it is apparent that these men are more than just a group who meet once a month at the exclusive Racquet Club, but rather represent a social system of corporate influentials related by business transactions. These lines of influence not only provide a social structure for promoting consensus in business affairs, but also in social policy decisions.

Table 5

CIVIC PROGRESS COMPANIES BY NUMBER OF EMPLOYEES

Size of Company	Number of Companies
4000+	8
2000–3999	4
1000–1999	7
500– 999	5
400– 499	1
300– 399	1
200– 299	1
100– 199	1
Less than 100	1
Total	29

Metropolitan St. Louis Large Employers, 1966, Chamber of Commerce of Metropolitan St. Louis.

Another aspect of influence channels entails officerships and board memberships of Civic Progress members in key agencies. Figure 1 represents the control Civic Progress has over selected agencies which play pivotal roles in community policy making. Control comes from individual members and from collective participation by Civic Progress in the activities of other agencies. All but the Civic Center Redevelopment Corporation are nonprofit agencies. The United Fund and The Arts and Education Council are the only two which conduct annual fund raising campaigns for member agencies. Both the Health and Welfare Council and Hospital Planning Commission are subservient to the United Fund because they are member agencies depending on it for basic support. Lines of influence flow from the UF to HWC in terms of control of program. However, the HPC is somewhat autonomous because of its funding from the City of St. Louis, St. Louis County, and the United States Public Health Service. Although both are under the same executive director, they maintain separate boards.

Table 8 gives a measure of board interpenetration of these seven planning agencies. Civic Progress members held a total of eighty-two board positions in these six agencies. Of the thirty officerships, they had fifteen.

Table 6

CIVIC PROGRESS MEMBERS, 1966

Name	Title	Company
J. A. Baer II	President	Stix, Baer & Fuller
August A. Busch, Jr.	Pres. & Bd. Ch.	Anheuser-Busch
David R. Calhoun, Jr.	Pres. & Bd. Ch.	St. Louis Union Trust
Maurice R. Chambers	Pres. & Bd. Ch.	Interco
Kenton R. Cravens	Board Chairman	Mercantile Trust
H. Reid Derrick	President	Laclede Gas
James M. Douglas	Attorney	
Irving Edison	President	Edison Bros. Stores
Thomas H. Eliot	Chancellor	Washington University
Preston Estep	Board Chairman	Bank of St. Louis
Clark R. Gamble	Board Chairman	Brown Shoe
Theodore Gamble	President	Pet Company
Russell E. Gardner, Jr.	Senior Partner	Reinholdt and Gardner
Richard A. Goodson	President	Southwestern Bell
W. Ashley Gray, Jr.	President	General Steel Industries
Joseph E. Griesedieck	President	Falstaff Brewing
Harry F. Harrington	Board Chairman	Boatmen's National Bank
James P. Hickok	Board Chairman	First National Bank
Ben F. Jackson	Board Chairman	Price Waterhouse & Co.
Aloys P. Kaufmann	President	Chamber of Commerce
Morton D. May	President	May Department Stores
J. W. McAfee	Bd. Ch. & CEO	Union Electric
Wm. A. McDonnell	Ch. of Fin. Com.	St. Louis-San Francisco R.R.
Frederic M. Peirce	President	General American Life Ins.
W. R. Persons	Ch. and CEO	Emerson Electric
The Rev. Paul C. Reinert, S.J.	President	St. Louis University
Raymond E. Rowland	President	Ralston Purina
Chas. H. Sommer	Pres. & CEO	Monsanto Co.
L. J. Sverdrup	Board Chairman	Sverdrup & Parcel & Assoc.
Clarence M. Turley, Sr.	President	Clarence M. Turley, Inc.
Nicholas P. Veeder	Bd. Ch. & Pres.	Granite City Steel Co.

Robert K. Sanford, "Civic Progress, Inc., Members Contribute in Team Pattern," *St. Louis Post-Dispatch*, November 1, 1966.

Table 7

TOP BUSINESSMEN BY DIRECTORSHIP ON CORPORATE BOARDS, 1966

	Boatmen's Bank	Mercantile Trust	First Nat'l Bank	St. Louis Union Trust	Union Electric	Southwestern Bell Telephone	Gen. Amer. Life Insur. Co.	Emerson Electric	Granite City Steel	Interco	General Motors	McDonnell Aircraft	Monsanto	Pet Milk	Ralston Purina	St. Joseph Lead	Universal Match	Anheuser-Busch
Baer, J. A. II		X																
Busch, August A., Jr.			X	X			X											X
Calhoun, David R., Jr.			X	X	X			X		X			X			X	X	X
Clark, Edwin M.		X				X	X											
Cravens, Kenton R.		X							X	X				X				
Derrick, H. Reid	X																	
Edison, Irving	X																	
Gamble, Theo. R.			X	X			X							X				
Hickok, James P.			X	X														
McAfee, J. W.			X	X	X	X	X				X							
McDonnell, Wm. A.			X		X		X					X						
Peirce, Fred W.							X											
Persons, W. R.								X										
Rowland, Raymond E.		X	X	X										X	X	X	X	X
Sommer, Chas. N.			X	X					X				X					X
Sverdrup, Leif J.															X		X	X

Poor's Register of Corporations, Directors and Executives, U.S. and Canada (New York: Standard and Poor's Corporation, 1966).

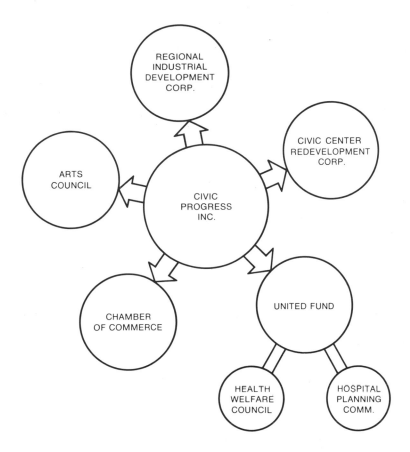

Fig. 1. Model of Private Organization of Social Welfare in the City of St. Louis.

Table 8

BOARD POSITIONS OF CIVIC PROGRESS MEMBERS IN
FUNDING AND PLANNING AGENCIES, 1966

Name	UF	HPC	AC	CC	RIDC	CCRC
Baer, J. A. II	x				x	
Busch, A. A., Jr.	x				x	x
Calhoun, David R., Jr.	x	x			x	x
Chambers, M. R.	o	x		x	x	
Cravens, Kenton R.	x	x		x		
Derrick, H. Reid	x	x		x		x
Douglas, James M.	x	x				
Edison, Irving	o			x		
Eliot, Thomas H.	x					
Estep, Preston	x					o
Gamble, Clark R.	x			x		
Gamble, T. R.	x				x	
Gardner, Russell E.	x	x				
Goodson, R. A.	x			x		
Gray, W. Ashley, Jr.	x			x	x	
Griesedieck, J. E.	x		x		o	
Harrington, H. D.	x	x		x		x
Hickok, James P.	x			x		o
Jackson, Ben F.	x	x		o		
Kaufmann, Aloys P.	x			o		o
McAfee, J. W.	x					x
McDonnell, W. A.	x			x	x	
May, Morton D.			o		o	o
Peirce, Frederic M.	o	x		o		
Persons, W. R.	x				x	
Reinert, Rev. Paul	x	x	x			
Rowland, Ray E.	x	x				x
Sommer, C. H.	x					
Sverdrup, L. J.	x			x		
Turley, Clarence M., Sr.	x				x	
Veeder, N. P.	x	o		x	o	
Totals	30	12	3	15	12	10

o = officer; x = board member.
Archives of Health and Welfare Council.

All but one of the Civic Progress members were on the board of the United Fund. Frederic Peirce was ex officio member of the board of directors of the Health and Welfare Council because of his being president of the United Fund. Otherwise, there was no representation from Civic Progress. Planning councils like the Health and Welfare Council are generally viewed as "departments" of united funds. Hence, board participation by chief executive officers of the largest corporations in HWC is not considered essential.

Besides being in the pivotal funding and planning agencies, Civic Progress members are on boards of the more prestigious agencies such as hospitals, Boy Scouts, and the YMCA. The Boy Scouts have seventeen officers on their board, and eight of these were from Civic Progress companies. Of these eight, six were regular members and one an emeritus member of Civic Progress. The central board of directors of the YMCA has ten officers, seven of whom worked for Civic Progress corporations. Two of these officers were in Civic Progress.

Not only are there superordinate and subordinate relations between corporations which operate in a more or less unorganized fashion, it is inferred from the preceding evidence regarding lines of influence that there is monopolistic control based on the existence of an agency, Civic Progress, which coordinates the largest locally based corporations. The fact that there are in St. Louis hundreds of little agencies in what appear to be organizational anarchy may be like chickens among the dancing elephants (Hacker, 1965:7). These different agencies depend on money from Civic Progress corporations. The existence of some relatively unrelated agencies may be functionally desirable in a corporate monopoly of social policy.

Robert Sanford, *St. Louis Post-Dispatch* feature writer, did a series on Civic Progress, Inc. (Sanford, 1966a, 1966b, 1966c, 1966d, 1966e). His conclusions point to this voluntary agency's being an executive committee of corporate interests. Their activity in politics has already been noted and recurs again and again as a theme. It is this way because that is the essence of a welfare system. It is total. Business is politics and politics is business. This is the case both domestically and internationally. The State here and abroad must often heel.

Officers of Civic Progress in 1966 were: David R. Calhoun, Jr., Chairman (formerly President); W. R. Persons, President; and Preston Estep, Treasurer. Persons is the CEO of Emerson Electric and its former head was Stuart Symington, now U.S. Senator. To argue, as do pluralists Banfield and Dahl, that businessmen tend not to be interested in politics is to lose crucial relationships in a big numbers game. Preston Estep, head of the General Bancshares Corp., a holding company for banks, the chief one of which is the Bank of St. Louis, is proof of how poor country boys can find a way to the top. He is not like the Buschs, Gambles, Danforths, noblemen by birth, who inherited ladders to the castle heights. He works well with Mayor Cervantes and has helped to get bond issues passed.

Although certain achievements were noted in the preceding chapter, the following takes Sanford's reporting of the affairs of Civic Progress and puts them into a chronological sequence.

1954

Civic Progress obtained an earnings tax to be paid by all persons employed in the city. It provided leadership to raise funds for the Pope Pius XII Memorial Library at St. Louis University and raised $65,000 to subsidize the faltering St. Louis Symphony.

1955

Civic Progress campaigned successfully for passage of $110 million public improvement bonds for the city, and also helped in the passing of the $39 million bond issue for St. Louis County. The Community Chest was reorganized into the United Fund by Civic Progress leadership. Of the forty-seven member executive committee of the United Fund, eighteen are Civic Progress members. Seven Civic Progress members have been Campaign Chairmen in the past twelve years.

1956

Civic Progress succeeded in a campaign to elect freeholders to draft a city charter.

1957

Civic Progress campaigned for but lost the election on the charter revisions proposed by the freeholders.

1958

Downtown, Inc. was set up by Civic Progress to give more stimulus for various programs to revitalize the downtown. Much change came about —new buildings, more parking lots, and new street lighting. Downtown, Inc. is responsible for Mansion House; it is currently sponsoring the national design competition for the Gateway Mall. It has generated interest in renewal of Laclede's Landing.

1959

Civic Progress campaigned and was successful in getting the earnings tax raised to 1 percent.

1960

Civic Progress led a successful campaign for sewers and related facilities.

1961

Civic Progress organized the Metropolitan St. Louis Hospital Planning Commission.

1962

Civic Progress backed a successful campaign for a $95 million Metropolitan Sewer bond issue; also supported municipal and school bond issues. Civic Progress worked for tax increases for the public library, zoo, and art museum. Through Downtown, Inc., Civic Progress obtained revision in the city building codes, making possible use of new materials in Mansion House. Civic Progress assisted in bringing about the Junior College District.

1963

Civic Progress formed the Regional Industrial Development Corporation in order to attract new industry to the St. Louis area. Five members of Civic Progress are on RIDC's executive committee. Civic Progress was influential in setting up the Arts and Education Council, and gives leadership to its annual campaign for funds.

1964

No new activities reported by Sanford.

1965

Civic Progress campaigned successfully for St. Louis County's $40,-699,000 bond issue for roads and incinerators. Civil rights was designated as the major community problem, and a committee was set up to monitor the racial situation.

1966

Civic Progress established Work Opportunities Unlimited as the businessman's approach to hard core unemployment. WOU finds jobs and makes placements. Members of Civic Progress contributed more than $10,000 and applied for money from the Bureau of Apprenticeship Training of the Department of Labor and the Office of Economic Opportunity. The Regional and Washington OEO Officers questioned the motive of Civic Progress, but Sanford said that strong support from Civic Progress settled their doubts. Also, during this year, Civic Progress contributed $50,000 for the campaign to pass the $79.3 million general obligation improvement bonds. Civic Progress had worked on the initial screening and paring down of the originally proposed $150 million bond issue.

Civic Progress, Sanford indicates, provides leadership and money for community projects. Chief executive officers are usually experienced organizers, and with the wealth of their corporations at their disposal, there is much they can do. Despite having only a part time staff person who is a public relations counselor, substantial resources are made available to the community. Originally Civic Progress planned to model itself after the Allegheny Conference on Community Development in Pittsburgh and have a sizeable staff. However, Civic Progress stayed small. Civic Progress members decided, Sanford says, that big organizations spend too much time going to meetings with little results. Civic Progress is able to act informally and more or less behind the scenes with a minimum of public discussion. These men have close business ties with each other, often meeting in board rooms of fellow Civic Progress members or at board meetings of other civic and social welfare agencies. He reports that St. Louis Union Trust Company had eleven Civic Progress members on its board, The First National Bank (affiliated with St. Louis Union Trust) had eleven, Mercantile Trust had five, and eleven of the fourteen member board of General American Life Insurance Co. were Civic Progress members (three being emeritus members). Some members are particularly active in each other's companies: Calhoun is a director in six firms besides his own; L. J. Sverdrup is on boards of four Civic Progress concerns. The general rule, Sanford comments, is for a Civic Progress member to be on the board of at least one other company represented on Civic Progress.

In order not to be accused of preempting too much, Sanford says they set up new agencies. Sanford, while failing to indicate the mechanisms of such preemption, probably had in mind the following examples. Downtown, Inc. and the Regional Industrial Development Corp. are spin-offs from Civic Progress. Civic Progress could have undertaken these but it would have required becoming a big organization open to accusations of empire building—trying to dominate St. Louis. Effective influence over the new agencies is obtained through interlocking directorates without the risk of unfavorable publicity. At the same time, Civic Progress could stay small and flexible. Although Sanford speaks of Civic Progress working through existing organizations, it would seem that usually these are the ones which Civic Progress itself has instituted. There are old line agencies in which members of Civic Progress are active such as the Boy Scouts and YMCA, which by their prestige legitimate activities of Civic Progress.

David R. Calhoun, Jr. is quoted by Sanford as denying there is a single power structure but that labor has its own power structure. Civic Progress does not initiate projects and does not want to dominate, says Calhoun. W. R. Persons characterized it as a senate for thinking about metropolitan problems. Groups wanting money, public backing or both, often seek a hearing before Civic Progress which meets monthly at the Racquet Club. Spokesmen for a cause describe what they want to do and then answer questions; after their departure, the proposal is evaluated by Civic Progress members who may decide to support or turn it down on the spot or to obtain consultation. Help may be only from individual members or Civic Progress may undertake the program as a special project. These men are not accustomed to failure, Sanford declares. There is esprit de corps in Civic Progress indicated by the required attendance at meetings. Also, no matter how busy they are, they are supposed to carry out their Civic Progress commitments. Although they delegate responsibility to their company executives, there are community jobs which only the chief executive officer can do in outstanding fashion.

Protestations notwithstanding, economic, political, civic and social welfare activities are apparently interrelated and interdependent. Social welfare or civic interests can be economically and politically beneficial to business elites. Powerlessness due to seeming fragmentation or multiplicity of power structures is probably illusory.

The following classifies and describes some of the operations of community leaders, drawing on evidence previously presented in this book plus additional findings: (1) philanthropy; (2) fund raising; (3) regulating fund drives; (4) agency making; (5) holding strategic positions; (6) community spokesmen, and (7) honored citizens.

Corporate influentials are much publicized for giving money to charities and educational institutions. August A. Busch, Jr. received the Boys' Club Man and Boy Award because he gave Busch Stadium to the Metropolitan St. Louis Boys' Club ("Nixon Presents Highest Honor—Boys' Club Cites Busch," *St. Louis Globe-Democrat,* November 29, 1966). Busch also gave $600,000 to St. Louis University for its 150th anniversary leadership drive ("Donated by Anheuser-Busch $600,000 to St. Louis U.," *St. Louis Post-Dispatch,* March 4, 1967). The chairman of this development project is Harry F. Harrington who, like Busch, is a member of Civic Progress. Chancellor Thomas H. Eliot of Washington University accepted a grant of $500,000 from Raymond E. Rowland, Board Chairman of Ralston Purina Company ("Ralston Pledges $500,000 to W. U.," *St. Louis Globe-Democrat,* May 8, 1967). The latter two are in Civic Progress. The money is pledged for the Medical School and is a contribution to the university's drive for $70 million by 1970. In the 1966–67 United Fund Campaign, James S. McDonnell is shown with his associates giving checks totaling $630,000 to W. Ashley Gray, Jr., United Fund campaign chairman ("McDonnell Donates $630,000, Largest in United Fund History," *St. Louis Globe-Democrat,* October 4, 1966). Of the $630,000, $490,000 was from the McDonnell Personnel Charity Trust and $140,000 as a corporate contribution. Although McDonnell is not in Civic Progress, Gray is.

The line between philanthropist and money-getter is thin. Influentials perform both roles. Fund raising on behalf of the United Fund and for the City of St. Louis in connection with bond issues has already been noted. They lend their efforts to other campaigns besides. Funds for the new General MacArthur Boy Scout Building were raised almost entirely by General Leif J. Sverdrup, a member of Civic Progress ("The General MacArthur Boy Scout Building," *St. Louis Globe-Democrat,* May 26, 1967). The St. Louis Symphony conducted a capital fund drive to raise $4 million, and Maurice R. Chambers, General Chairman of the drive and in Civic Progress, reported that pledges of $3,650,000 had been received

("The Symphony Nears Its Goal," *St. Louis Post-Dispatch* editorial, June 29, 1967). The Rose Ball to help the St. Louis Association for Retarded Children had Mrs. Joseph P. Kennedy as guest of honor ("Everything Came Up Roses," *St. Louis Post-Dispatch,* June 11, 1967). David R. Calhoun, Jr., Civic Progress member, introduced her. Preston Estep, John L. Wilson and Clarence L. Turley, Sr. are co-chairmen of the drive to raise $2 million to reconstruct the Spanish International Pavilion in St. Louis ("Ground Broken for Rebuilding of the Pavilion," *St. Louis Post-Dispatch,* June 30, 1967). Estep and Turley are in Civic Progress; the former has been chairman of City Bond campaigns, and the latter has large real estate interests in downtown St. Louis. Wilson is chief executive officer of Universal Match Corporation. Even the Teamsters Joint Council 13 emulates the business influentials in having an annual charity show ("$257,000 Charity Gifts by Teamsters," *St. Louis Post-Dispatch,* June 17, 1967). They allocated $257,000 to local charities, national and international organizations. It is of interest that two of the favored social agencies (Boy Scouts and the Metropolitan St. Louis Boys' Clubs) of the business influentials were included in the list of the Teamster beneficiaries.

Business leaders regulate the fund drives of various agencies through the Charities Committee of the Chamber of Commerce, and the Metropolitan St. Louis Capital Fund Review Board which is an outgrowth of the former. The Statement of Principle of the Charities Committee (1966:6) says it is "to provide . . . a list of approved, local charitable organizations. The Charities Committee does not attempt to tell anyone if he should or should not give. It merely supplies the facts." The Capital Fund Review Board is to reduce duplication in campaigns, and endorsement is to assure givers that the cause is legitimate ("Board to Screen Capital Fund Drives," *St. Louis Post-Dispatch,* September 15, 1966). Civic Progress members had a hand in the formation of the latter, but only their endorsement of the Board was made public ("Fund Review Backed by Civic Group," *St. Louis Globe-Democrat,* November 30, 1966). They are in the unique position of being able through individual members to start something and then to have the organization formally back them as though the support were independent.

They form new agencies which can carry out the objectives of the business executives. Civic Progress, the United Fund of Greater St. Louis, Downtown, Inc., Regional Industrial Development Corporation, Hospital Planning Commission of Metropolitan St. Louis, Work Opportunities Unlimited, and the Herbert Hoover Boy's Club are some of the agencies which influentials, particularly those in Civic Progress have promoted.

They are elected or named to important posts. Richard S. Jones, senior Vice President of corporate development of Pet, was elected President of the Health and Welfare Council ("Welfare Based on Need Is Urged," *St. Louis Post-Dispatch,* June 2, 1967). Alvin Tolin, Controller of

Ralston Purina Co. was elected, and Gerard K. Sandweg of Mercantile Trust was reelected to the board of HWC. Although these are men of lesser influence in their corporations, the chief executive officers of these same corporations are in Civic Progress. R. Ray Shockley was elected trustee of Deaconess Hospital; Shockley is Vice President and General Manager of Southwestern Bell Telephone Co. ("Shockley Trustee of Deaconness," *St. Louis Globe-Democrat,* April 7, 1967). Nicholas P. Veeder, Board Chairman of Granite City Steel was named Chairman of the Governor's Council on Alcoholism ("6 St. Louisans on Alcoholism Advisory Council," *St. Louis Globe-Democrat,* May 5, 1967). David R. Calhoun, Jr. and J. W. McAfee were appointed co-chairmen of the United Fund's Chapter Plan Committee ("Calhoun, McAfee to Aid Fund as Co-Chairmen," *St. Louis Globe-Democrat,* May 5, 1967). The Very Rev. Paul C. Reinert, S.J., President of St. Louis University, announced that David R. Calhoun, Jr. and William A. McDonnell have been named co-chairmen of St. Louis University sesquicentennial ("Two Executives Named to Head St. Louis U. Sesquicentennial," *St. Louis Post-Dispatch,* October 23, 1966). He was quoted as saying: "Both have distinguished themselves in this community, not only in business but also in civic affairs. Both are members of Civic Progress, Inc., which has spearheaded the rebirth of downtown St. Louis." Father Reinert is also a member of Civic Progress.

Business influentials speak out on community issues. The *Globe-Democrat* asked sixteen area leaders what should be given priority in 1967, and seven of these are in Civic Progress ("Can St. Louis Keep Rolling?" *St. Louis Globe-Democrat,* December 31, 1966; January 1, 1967). Business leaders are also extollers of the American way (William K. Wyant, "McDonnell has High Praise for American Way," *St. Louis Post-Dispatch,* May 25, 1967). In addition, their voices are heard in high places asking for special favors. August A. Busch, Jr. and the Very Rev. Paul C. Reinert, S.J. along with some other citizens of St. Louis asked for Finnegan's pardon ("Presidential Pardon Given to James Finnegan," *St. Louis Post-Dispatch,* May 3, 1967). Finnegan was convicted for misconduct in office while serving as collector of internal revenue in St. Louis.

They are often honored. There are two major awards given each year: the St. Louis Award and the *Globe-Democrat's* Man of the Year. Richard Amberg, Publisher of the *Globe-Democrat,* received the St. Louis Award for his civic activities in 1966 ("For Service to Community Richard H. Amberg Wins the St. Louis Award," *St. Louis Globe-Democrat,* December 9, 1966). Since 1953, six of the winners of this award have been members of Civic Progress. In the same year, the Man of the Year Award went to Charles Allen Thomas, Chairman of the Board of Monsanto Company.

> He was the choice of a selections committee composed of the 11 [sic] previous winners of the Man of the Year Award. The members are

David R. Calhoun, Jr., President of the St. Louis Union Trust Company; Maj. Gen. Leif J. Sverdrup, prominent engineer and head of the firm Sverdrup & Parcel, Inc.; Ethan A. H. Shepley, lawyer and former Chancellor of Washington University; United States Senator Stuart Symington of Missouri (formerly President of Emerson Electric); Morton D. May, President of the May Department Stores Company; Congressman Thomas B. Curtis of the Second Congressional District; August A. Busch, Jr., President of Anheuser-Busch, Inc.; Edwin M. Clark, retired former President of Southwestern Bell Telephone Company; the Very Rev. Paul C. Reinert, S.J., President of St. Louis University; H. Sam Priest, President and executive officer of the Automobile Club of Missouri, and James P. Hickok, Chairman of the Board and chief executive officer of the First National Bank in St. Louis. Richard H. Amberg, Publisher of the *Globe-Democrat,* is also a member of the committee ("Charles Allen Thomas, The Globe-Democrat's Man of the Year," *Globe-Democrat Sunday Magazine,* December 25, 1966, p. 6).

All but Symington, Curtis, Priest, and Amberg are in Civic Progress either as active or emeritus members. The Distinguished Executive Award went to General Leif J. Sverdrup, in the above-mentioned list of civic notables ("Sverdrup to Receive Top Executive Award," *St. Louis Globe-Democrat,* April 4, 1967). The following month he received the Citizen of the Year Award from the Third District of Loyal Order of Moose ("General Sverdrup Receives Citizen of the Year Award," *St. Louis Globe-Democrat,* May 29, 1967). Morton D. May, also a former Man of the Year, was so acclaimed the preceding year. In June General Sverdrup was again honored, this time by the Man of the Year Award given by the Kiwanis Club of South St. Louis ("L. J. Sverdrup Gets Man of Year Award," *St. Louis Post-Dispatch,* June 10, 1967). James McDonnell received the 1966 Junior Achievement "Free Enterprise Award" ("James McDonnell to Receive Award," *St. Louis Globe-Democrat,* May 3, 1967). These business luminaries stand out above the ordinary men. They are not like movie celebrities or some of the national politicians, but they are known by the business community because they are continually in the paper and because of their economic influence. They employ many St. Louisans, build the tall buildings to house their corporations (both Pet and Laclede Gas are doing so in the downtown area), erect the civic edifices and monuments. They literally carve out the shape of the city.

A UNION VIEW

Ernest Calloway (1964), associate research director for the central conference of Teamsters in St. Louis, did an outstanding community service in 1964 by his assessment of the city's power structure. He found that economic elites are the omnipotent rulers. In addition to the economy, they control church, politics, the press, and social welfare. Calloway outlines five classes of dominants: (1) industrial; (2) banking and investment; (3) mercantile and commercial; (4) old family and institutional trust hierarchies, and (5) foreign investment and managerial bloc. The hard core of the power structure consists of informal and formal relationships of indigenous mercantile, banking and industrial groups. He lists twenty-five chief executive officers of corporations with assets representing $10 billion and employing about 325,000 of whom 100,000 are St. Louisans.

Table 9

THE INDIGENOUS ECONOMIC FRONT OF ST. LOUIS

Name	Firm	Assets in Thousands
Edwin M. Clark	Southwestern Bell	$2,900,000
David R. Calhoun, Jr.	St. Louis Union Trust Co.	25,000
C. A. Thomas	Monsanto Chemical	1,325,000
D. B. Jenks	Missouri Pacific	1,160,000
J. Wesley McAfee	Union Electric	920,000
Sidney Maestre	Mercantile Trust Co.	791,000
J. P. Hickok	First National Bank	738,500
Morton D. May	May Department Stores	433,500
Donald Danforth	Ralston Purina Co.	266,500
H. F. Harrington	Boatmen's National Bank	253,500
N. P. Veeder	Granite City Steel Co.	214,500
August A. Busch, Jr.	Anheuser-Busch, Inc.	204,000
Henry H. Rand	International Shoe Co.	199,500
H. R. Derrick	Laclede Gas Company	173,000
J. S. McDonnell	McDonnell Aircraft	149,000
Clark R. Gamble	Brown Shoe Company	142,500
W. R. Persons	Emerson Electric Co.	112,000
John L. Wilson	Universal Match Company	63,000
Irving Edison	Edison Brothers Stores	61,000
C. P. Whitehead	General Steel Industries	72,000
George W. Brown	Wagner Electric Co.	71,500
W. M. Akin	Laclede Steel Company	51,000
Ethan H. Shepley	Washington University Board	*
L. J. Sverdrup	Sverdrup and Parcel and Assoc.	*
Frederic M. Pierce	General American Life Ins.	*

* Not available.

Ernest Calloway, "The Nature and Flow of Economic Power," *Missouri Teamster* (February 21, 1964), p. 2.

Table 10

THE MOST INFLUENTIAL POLITICAL DECISION MAKERS
IN THE CITY, 1964

Name	Primary Affiliation	Interest
J. Wesley McAfee	Union Electric Co.	New Industry & Investment
David R. Calhoun, Jr.	St. Louis Union Trust	New Industry & Investment
Mayor Raymond Tucker	Mayor of St. Louis	Arts of Accommodation
Aloys Kaufmann	St. Louis Chamber of Commerce	Business vitality
Sam Priest	Board of Police Commissioners	Law and order
J. S. McClelland	St. Louis Board of Education	Financing public education
Charles Farris	Urban Renewal	Financing slum clearance
Joseph Pulitzer	*Post-Dispatch*	View from the top
Richard Amberg*	*Globe-Democrat*	View from the top
Sidney Salomon	Insurance broker	Political balancing
Alfred Fleishman	Public relations advertising	Political balancing
Tom Guilfoil	Bi-State Dev. Agency	Political balancing
John J. Dwyer*	Chm. Democratic Central Committee	Vote-getting
Louis G. Berra	Democratic Committeeman	Vote-getting
John J. Lawler	Democratic Committeeman	Vote-getting
Anton Sestric	Former Democratic Committeeman	Vote-getting
Frederic Weathers	Democratic Committeeman	Vote-getting
Al Cervantes	Former Pres., Board of Aldermen	Shift in political balances
Morris Shenker	Attorney	Shift in political balances
Mark Holloran	National Democratic Committee	Posture of Dem. Party
Harold J. Gibbons	Teamsters Joint Council	Labor's interest
Larry Connors	Machinist District Council	Labor's interest
Joseph Clark	St. Louis Labor Council	Labor's interest
Cardinal J. Ritter*	Catholic Church	Moral and social posture
Mrs. L. H. Pincus	St. Louis League of Women Voters	Community posture

* Deceased.

Ernest Calloway, "St. Louis Power Structure Part II Politics—A Power-Balancing Instrument," *Missouri Teamster* (February 21, 1964), p. 3.

Politics is more of an open game, according to Calloway, because political elites must make accommodations to competing groups. Nevertheless, he says that the political structure is subordinate to the economic omnipotents. As with the business leaders, he lists twenty-five of the most influential political decision makers. The chief executive officers of St. Louis Union Trust Company and Union Electric are both listed in Tables 9 and 10. The head of the Chamber of Commerce was formerly Mayor of St. Louis from 1943–1949. He was also the last Republican mayor. The public relations advisor named as one of the political leaders is closely tied in with the economic power structure. In spite of uncertainties in politics, the ruling elite is solidified by the one-party system. Although philosophically Republican, Calloway says, the business community pursues its political interests through the Democratic Party on the local level.

Church, press, political organizations, judiciary, police and charities are specified as power balancing institutions by Calloway. They legitimate the economic omnipotents by manufacturing illusions; they smooth the road for policymakers by sustaining and refining major actions. Table 11 lists the institutions which Calloway noted. Their customary functions obscure the justification and sanctification which they give to economic influentials. Illusions of community well being, good image, nobility, superiority, humanity, morality and virtue, equality, knowledge, and justice sanction those who control policy. Calloway asserts that bigness is the

Table 11

POWER BALANCING INSTITUTIONS IN ST. LOUIS

The Democratic Party of St. Louis
The Republican Party of St. Louis
The St. Louis Post-Dispatch
The St. Louis Globe-Democrat
Network Radio and Television Outlets
The St. Louis Board of Police Commissioners
The St. Louis Board of Education
The United Fund
Major Philanthropic Foundations
Metropolitan Church Federation
Catholic Archdiocese
St. Louis Bar Association
St. Louis County Medical Society
Municipal, state and federal court system
Private and state institutions of higher learning

Ernest Calloway, "Social Illusion: The Institutional Role," *Missouri Teamster* (February 21, 1964), p. 6.

source of illusions: e.g. Big Civic Pride makes the illusion of Superiority and Nobility; Big Philanthropy, Humanity. The "power balancers" depend upon big business for survival, rationalizing their double role as being consistent with their own interests and accept economic power as synonomous with right. The daily newspapers cast the large corporations and their executives in the hero roles; labor and Negroes are those who cause problems.

It is noteworthy that in discussion of counter-pressures on economic dominants Calloway mentions only labor unions and Negroes. What has been earlier described as a multilateral arena—city politics—is considered as part of the power balancing mechanism. A one party system has little to push against.

St. Louis is shown to be a strong union center in terms of the ratio of union membership to the labor force. It is third highest among the big cities. Calloway writes that approximately fifty percent of the union members in St. Louis belong to three unions: Teamsters Joint Council No. 13, 42,000; Machinist District Council No. 9, 35,000; and St. Louis Building and Construction Trades Council with 27,000. Unionism is faced, he indicates, with a struggle for survival. He describes the attack on the Teamsters in St. Louis led by a Federal Grand Jury. Local 688 was investigated and its chief executive jailed for failure to show the union books. The "lynch mob," Calloway comments, was joined by the press. Although such narrative is hardly contrived to impress the reader that unions are a strong counter-pressure force, the Teamsters in St. Louis are known for their political action. He says their greatest influence is on politics and weakest on churches and civic pride. Labor's token voice in the United Fund is described. The economic notables decide how much money will be raised and who will be included in the benefactions.

He pictures the Negro community as not being homogeneous and Negroes having to live in a hostile environment. He states that Negroes constitute about one-third of the city's voters, and forty percent of the hard-core Democratic vote; also, that of Northern cities, the Negro percentage population in St. Louis is the third highest. Sensitive to the Negro ground swell of the early 1960s, and correct about the Negro potential for forcing economic dominants to make concessions, Calloway is also factual in seeing the Negro leadership usually supporting the power bloc. The Negro affirmative voting for the revitalization of the city's core is a case in point. Furthermore, and unmentioned by Calloway, the Negro influentials are susceptible to being preempted by the white influentials. Gains are made but at the expense of compromises with the white elites. Negroes, he declares, duplicate the foibles of the whites, and are consequently attracted by power and wealth of which white economic rulers are the gatekeepers. He perceives the Negro boring from within, following the proverbial philosophy of cooptation: "If you can't beat 'em, join 'em."

The St. Louis Charter fight of the late fifties, Calloway reports, was an attempt by influentials to dispossess Negroes of political power. (This involved, among other things, having aldermen elected at large, and was viewed by Negro leaders as a scheme to reduce their representation on the board of aldermen. Due to the pattern of segregation, Negroes were assured of representation, but if the Charter plan had been adopted, their position would have been considerably weakened.) Negro politicians, with the help of the Teamsters (which he does not say), successfully confronted the power group and maintained control of the Negro political power of the city. Containment policies no longer worked as shown by the Charter defeat, he comments. In addition public accommodation and fair employment laws were passed locally, he contends, because of Negro pressure; but the business elites still set the tempo of change.

The profit stimulus, Calloway notes, fosters slums. Slum living is caused by pressure of economic influentials and their institutional legitimators to keep wages low and living marginal. Slums exist because of the short-term profits made from them. Low capital investment, minimal upkeep costs, with decreasing taxes add up to substantial margins of profit. Moreover, slum properties are less expensive to the city short of funds because the money can be funneled into those parts of the city having greater vote potential than the slums. The city, however, becomes less attractive for business and for residential use; but federal urban renewal programs are available bringing millions of dollars to the city for redevelopment, also profitable for the "brick and mortar" group which he identifies:

Table 12

THE BIG "BRICK AND MORTAR" CIVIC PRIDE FRONT IN ST. LOUIS

General American Life Insurance Co.
Southwestern Bell Telephone Co.
St. Louis Chamber of Commerce
St. Louis Real Estate Board
Civic Progress, Inc.
First National Bank
Mercantile Trust Company
Boatmen's National Bank
St. Louis Union Trust Co.
St. Louis Development Corp.
Greater St. Louis Savings and Loan League
Washington University
St. Louis University

Ernest Calloway, "Flow of Power: Feeding & Care of Slums," *Missouri Teamster* (February 21, 1964), p. 5.

Thus, the latent function of physical deterioration seems to be freeing the land for more profitable uses.

In summing up, he says the major purpose of power is to legitimate and sanctify itself. In St. Louis, the basic concern of the economically powerful is how to hold and expand their dominance. He comments on how this was formerly accomplished by developing a closed economy, but today the emphasis has changed to attracting outside industry. Calloway, except for references to the urban renewal programs, did not elaborate the federal role in business and welfare. He seems to back away from his main thesis when he adds the qualification that crucial decision making is manysided. Outright oligarchy is rejected by Calloway because power balancing mechanisms are not wholly subservient. As important as labor and blacks are in providing counter-pressure, by his own evidence, they seem to be subdominants. Both have their interests to protect, and in this respect how different are they from the power balancing institutions? Since wages come from the corporation, unions are certainly dependent upon the corporations and labor subservient by definition. The union leaders are business executives too.

CONCLUSIONS

A comparison of Calloway's and Sanford's findings shows that twenty-three of the twenty-five on Calloway's list of economic influentials either are or were members of Civic Progress. McAfee and Kaufmann, listed as political decision makers by Calloway, are also on Civic Progress. Domination of the United Fund by business influentials is supported by both writers. *Weakness of labor and blacks is perhaps shown by their not being in Civic Progress during the period of this study.*

Calloway and Sanford find business influentials major deciders of social policy in St. Louis. Calloway utilizes economic and sociological analysis in demonstrating his position. A rather unsophisticated descriptive sociology, used journalistically by Sanford, highlights the preeminence of Civic Progress, Inc.

Calloway's analysis indicates that the church, press, political organizations, and others, legitimate corporate elites while opposing the corporate influentials are unions and blacks. Calloway did not report on the interrelationship of specific organizations as such. In fact, Civic Progress gets put in the Big "Brick and Mortar" group in Table 12. Sadly enough, he contributes to pluralist arguments by failing to deal with Civic Progress' interpenetration of the important community agencies. His union commitments perhaps cause him to ignore the close interlocking of business corporations and the unions which inherently exists because of the latter's acceptance of the former as superordinate. Similarly, he overlooks the trend for government to define a lack of consensus between the corporation and the union as violating the public interest.

Therefore, it is clear that private influentials control the interagency organization of social policy, and that guiding this process is Civic Progress, Inc. Sources and lines of influence are developed for the corporation executives who are members of Civic Progress. Their relationships with policy setting agencies are unambiguous. The control of these agencies is the way in which the Civic Progress executives have influence over money and men.

Chapter 6

The United Way

This is primarily the story of the Health and Welfare Council of Metropolitan St. Louis, a captive agency of the United Fund and Civic Progress, Inc. Four periods stand out in the Council's history: (1) Pre-Civic Progress; (2) Civic Progress Take-Over of the Community Chest; (3) Council Enslavement, and (4) Council Regains Power. It is a case history of pluralistic failure and oligarchial success. Even in its pluralistic zenith, the Council was not populist or mass based, but like the Community Chest, it was an agency's agency. The Council, like the Chest, buckled under to business elites, to large corporate givers. This was not unique to St. Louis; it happened throughout the United States. Philanthropy is big business and big business is the welfare system. The issues are often thought too important for government or little people to decide.

The reason for Civic Progress' interest was the large sums of money expended for health and welfare services. Since policy is controlled through funding, it is important to understand the sources of money. Figure 2 indicates that most money for health, welfare, and recreation is from private sources[5] (Combination of sectarian and nonsectarian funds: $81.9 + $39.9 = $121.8 million). The $121.8 million exceeds the public funds by about $33 million. However, in the private sector, the largest amount of income is being paid as fees for hospital services. This would suggest that a major private agency sector of St. Louis is outside the funding control of the large corporations but this is not the case. Civic Progress' control of the hospital planning field through the Hospital Planning Commission puts them in juxtaposition with the hospital field.

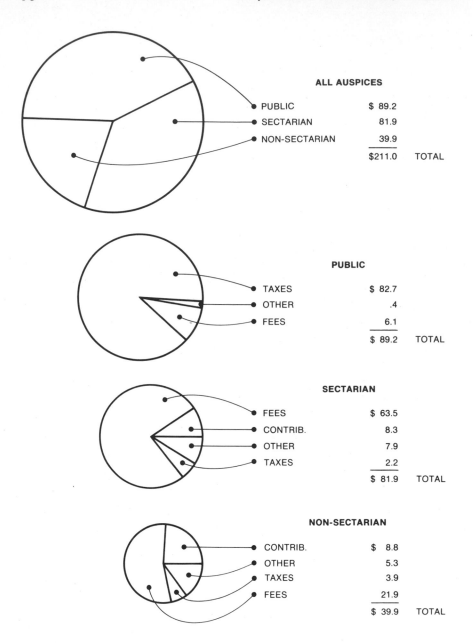

ALL AUSPICES

PUBLIC	$ 89.2
SECTARIAN	81.9
NON-SECTARIAN	39.9
$211.0	TOTAL

PUBLIC

TAXES	$ 82.7
OTHER	.4
FEES	6.1
$ 89.2	TOTAL

SECTARIAN

FEES	$ 63.5
CONTRIB.	8.3
OTHER	7.9
TAXES	2.2
$ 81.9	TOTAL

NON-SECTARIAN

CONTRIB.	$ 8.8
OTHER	5.3
TAXES	3.9
FEES	21.9
$ 39.9	TOTAL

Fig. 2. 1965 Health and Welfare Agency Income by Auspices and Sources, St. Louis (in millions of dollars).

PRE-CIVIC PROGRESS

The Council began in 1911 as the Central Council of Social Agencies. Twenty-eight agencies set it up agreeing to pay token dues. Its first head also served as secretary for the charities committee of the Chamber of Commerce. Ten years later the Central Council was changed to the Community Council with seventy-eight member agencies. The new Council was oriented to major problems such as pure milk, housing, and care of city institution inmates. The commitment to professional staffing became an issue then. Through the efforts of the Community Council, the Community Fund was formed in 1922 which today is the United Fund.

In 1937, the Council was reconstituted again as the Social Planning Council of St. Louis and St. Louis County. It developed a departmental and divisional organization. Various community agencies belonged to the different divisions of the Council.

CIVIC PROGRESS TAKE-OVER

The next transformation of the Council was in 1958 when it became the Health and Welfare Council. The change in the name is significant from the standpoint of indicating a somewhat more delimited role. Civic Progress had come into being four years before and had set up the United Fund in 1955 which superseded the Community Chest. The United Fund brought in more of the "top top" corporate leadership. Before continuing with the fate of the Council, UF development needs accounting.

Although there are different campaigns to raise money in St. Louis, the major effort is the United Fund which annually solicits money for agency services in the City of St. Louis, St. Louis County, and St. Charles County. Its purpose is stated as follows in the Articles of Incorporation:

> To raise funds and finance, in part, or in the entirety, local, state or national charitable, benevolent, eleemosynary, social, welfare, non-profit health and philanthropic organizations. To conduct fund-raising campaigns for such organizations, and to advise and acquaint the public of their objectives and purposes, their use of funds and such other information as will assist the public (*Information,* United Fund of Greater St. Louis, Inc., 1967:2).

In 1955 the United Fund was incorporated by the Rt. Rev. Msgr. John J. Butler, James M. Douglas, J. W. McAfee, and William A. Webb. Msgr. Butler was head of Catholic Charities; Webb the executive of the Central Trades and Labor Union, AF of L; Douglas, a former judge, and then an attorney, and McAfee the chief executive officer of the Union Electric Company. Both Douglas and McAfee are still in Civic Progress. The first board of directors of the United Fund had twenty-four members of whom fifteen were in Civic Progress.

The United Fund differs from the Community Chest organization in the following ways: (a) It is more inclusive. The American Red Cross is involved on a contractual basis, and attempts were made locally and nationally to include all major campaigns such as heart, cancer, tuberculosis and others. Opposition of the national organizations was largely responsible for the prevention of this inclusion. (b) The largest corporations took control of the fund raising and money allocating structure of the community and agreed that the United Fund would be given exclusive rights to in-plant solicitation and payroll deductions. (c) Agency representation in any formal sense (e.g., board of directors, executive committee, or budget committees) was eliminated. This was a development which took place in other cities in the United States and is referred to as the "revolt of the givers." To be more precise, it should be designated as the revolt of the "money getters."

The bylaws provide for 124 directors who select an executive committee composed of the officers of the UF plus no less than fifteen members from the board (United Fund of Greater St. Louis, 1967: Bylaws, Articles II and IV). The executive committee has full authority to act for the board. Besides setting policy, and supervising the activities of the corporation through the executive vice president (who is the administrative officer), they make the final determination of the allotments to the member agencies. The bylaws make no provision for agency representation on the board of directors, the executive committee, or any other structure of the corporation.

The Agency Relations Cabinet, which is appointed by the executive committee, recommends the allocation of funds, and releases the agencies to make sure they conform to the requirements which have been set for United Fund membership. The Agency Relations Cabinet consists of three officers and nineteen members. Two of the three officers are executives in corporations represented in Civic Progress, and four other executives from corporations in Civic Progress are in the cabinet. It

> . . . has the responsibility to consider the community health and welfare program as a whole and to relate each agency's service to that of every other in order to achieve a balanced community program to the greatest degree possible. . . . The distribution of United Fund monies must recognize and preserve the right and responsibility of the agencies to manage their own operations except by becoming or continuing to be participants in the United Fund as a joint enterprise and sharing in the proceeds of the United Fund campaigns, the

agencies voluntarily limit their freedom of action in the interest of the common good of all United Fund agencies and of the community as a whole (United Fund of Greater St. Louis, 1967: section on budgeting: 1).

Concerning agency autonomy: "It is United Fund policy to allow each agency as much autonomy as is consistent with efficiency and effectiveness and sound fiscal responsibility" (United Fund of Greater St. Louis, 1967:4). The Agency Relations Cabinet and staff of the United Fund, who work with this cabinet, regulate and supervise the agencies in terms of the following: (a) Budgeting of funds within the agency, budgets which include the sources of funding must be submitted for approval; (b) A salary and position classification plan was sct up, new and existing personnel must be fitted into salary ranges defined by the plan; (c) The program of the agency is reviewed by the Health and Welfare Council periodically at the request of the United Fund; and (d) Capital expenditures must be approved through the Agency Relations Cabinet.

Another facet of private control is the campaign itself. The campaign purpose is to raise money for allocations to member agencies. Concerning the organization of the campaign, the following is stated:

> The Campaign Chairman serves from the time of his appointment, usually in January, until the termination of the campaign, the end of October. He is assisted throughout the campaign by top level community leaders, each of whom accepts responsibility for a part of the total campaign effort. The St. Louis United Fund, the past five years has had the benefit of outstanding support from all media and from pace-setting labor support that is the envy of the nation. An Audit Committee is made up of volunteer auditors who devote their time during October to an actual audit of campaign returns as they are turned into United Fund headquarters. Campaign costs are kept at a minimum by securing Loaned Executives from firms and organizations throughout the campaign area. Nearly one hundred such Loaned Executives serve full-time for an eight-week period, starting the day after Labor Day in each campaign (United Fund of Greater St. Louis, 1967: Campaign Section 1: 1).

It is stated concerning fair share contribution: "The suggested Fair Share giving scale for individuals paid an hourly rate is six-tenths of one percent, which is the equivalent of one hour's pay per month for twelve months. For salaried persons, the scale is graduated upward and is proportionate to income" (United Fund of Greater St. Louis, 1967).

Labor unions are represented on the board and executive committee of the United Fund. In addition there are two full time staff members who are labor representatives from the American Federation of Labor and the Congress of Industrial Organizations (AFL-CIO). These have been re-

Table 13

FUND RAISING RECORD OF THE UNITED FUND AND ITS PREDECESSOR ORGANIZATIONS

	President	Campaign Chairman	Goal	Amount Pledged	Percent Raised
Community Fund					
1922–1923	W. Frank Carter	H. Johnston	$ 1,081,684	$ 1,125,954	104.1
1923–1924	Daniel N. Kirby	Thomas N. Dysart	1,495,837	1,439,247	96.2
1924–1925	Daniel L. Catlin	B. H. Lang	1,751,521	1,441,029	82.3
1925–1926	L. Wade Childress	Warren C. Flynn	1,848,768	1,644,352	88.9
1926–1927	Erastus Wells	Alfred Fairbank	1,848,727	1,712,253	92.6
1927–1928	Col. Albert Perkins	G.·M. Berry	1,850,000	1,748,037	94.5
1928–1929	Col. Albert Perkins	George M. Berry	1,902,272	1,725,018	90.7
1929–1930	Ethan A. H. Shepley	Sidney Maestre	2,092,935	2,002,413	95.7
1930–1931	Sidney Maestre	Gale Johnston	2,205,788	2,200,311	99.8
1931–1932	Sidney Maestre	Ethan A. H. Shepley	3,009,857	2,458,500	81.7
United Relief					
1932–1933	Charles Nagel	Frank O. Watts	2,850,145	2,372,048	83.2
1933–1934	Charles Nagel	Arnold G. Stifel	2,794,899	2,212,027	79.1
1934–1935	Charles Nagel	Leo G. Fuller	2,708,000	2,282,489	84.3
1935–1936	Oliver Richards	Fred C. English	2,888,000	2,184,590	75.6
United Charities					
1936–1937	Oliver Richards	Fred C. English	2,610,000	2,125,305	81.4
1937–1938	Oliver Richards	Henry W. Kiel	2,550,000	2,185,671	85.7
1938–1939	Oliver Richards	Henry W. Kiel	2,495,170	2,068,017	82.9
1939–1940	Oliver Richards	J. L. Ford	2,200,000	2,200,348	100.0
1940–1941	Ethan A. H. Shepley	Samuel D. Conant	2,254,493	2,158,906	95.8
1941–1942	Joseph Desloge	Benjamin M. Loeb	2,150,000	2,184,245	101.6

War Chest

Period					
1942–1943	Frank C. Rand	J. W. McAfee	$ 4,850,000	$ 5,201,605	107.2
1943–1944	Frank C. Rand	Benjamin M. Loeb	5,265,000	5,276,210	100.2
1944–1945	Frank C. Rand	Stratford Lee Morton	5,265,000	5,317,904	101.0
1945–1946	Frank C. Rand	Cyrus C. Willmore	5,265,000	4,836,199	91.9

Community Chest

Period					
1946–1947	Ethan A. H. Shepley	Henry Hitchcock	4,730,000	4,335,650	91.7
1947–1948	Ethan A. H. Shepley	Wm. A. McDonnell	4,785,000	4,405,959	92.1
1948–1949	Benjamin M. Loeb	Howard F. Baer	4,785,000	4,694,167	98.1
1949–1950	Benjamin M. Loeb	James P. Hickok	4,785,000	4,539,858	94.9
1950–1951	Wm. M. Rand	Donald Danforth	5,150,625	5,154,123	100.1
1951–1952	Wm. B. McMillan	Col. Clark Hungerford	5,519,700	5,523,123	100.1
1952–1953	William Charles	Louis A. Hager, Jr.	5,553,000	5,362,103	96.6
1953–1954	Robert H. Mayer	{ Norman J. George / Joseph H. Vatterott	6,082,592	5,784,444	95.1
1954–1955	F. Wendell Huntington	Chapin S. Newhard	6,100,000	5,803,960	95.1

United Fund

Period					
1955–1956	Donald Danforth	William A. McDonnell	8,245,000	8,245,000	100.0
1956–1957	Kenton R. Cravens	Charles P. Whitehead	8,250,000	8,250,000	100.0
1957–1958	Kenton R. Cravens	Hugh A. Logan	8,598,000	8,161,000	94.9
1958–1959	Samuel D. Conant	Felix N. Williams	8,598,000	8,061,000	93.8
1959–1960	John H. Hayward	Harry W. Chesley, Jr.	8,600,000	8,600,000	100.0
1960–1961	Ben F. Jackson	B. B. Culver, Jr.	8,600,000	8,600,000	100.0
1961–1962	B. B. Culver, Jr.	Ethan A. H. Shepley, Jr.	9,050,000	8,733,648	96.5
1962–1963	Edwin M. Clark	Raymond E. Rowland	9,200,000	9,477,948	103.2
1963–1964	Charles A. Thomas	Harry F. Harrington	9,200,000	9,740,520	105.9
1964–1965	C. Powell Whitehead	Frederic M. Peirce	9,500,000	10,152,963	106.9
1965–1966	James P. Hickok	M. R. Chambers	10,000,000	10,851,922	108.5
1966–1967	Frederic M. Peirce	W. Ashley Gray, Jr.	10,000,000	11,328,547	107.9

Totals (of 45 campaigns, 17 have met or exceeded their quotas): $221,005,013 | $215,908,613

Information for Members of the Board of Directors and Executive Committee of the United Fund of Greater St. Louis, Inc., 1967.

ferred to as sops for the labor unions. It is indicative of labor's subordinate status to the corporation that they would be given staff positions in the UF since those who are in key positions in the campaign organization are volunteer executives from large corporations.

In the twelve campaigns under the United Fund, there were only two years when the goals were not met. The campaign chairmen those years were not Civic Progress members, but in the successful years, the campaign chairmen were in Civic Progress. The campaigns of the predecessor agencies as shown in the same table were not as successful in meeting goals. Among the thirty-three drives preceding the establishment of the UF, only eight were victorious. Campaigns locally in the past twelve years have raised $110,202,548 and it is reported that over 40,000 volunteers help in each campaign (United Fund of Greater St. Louis, 1967). The goals have been larger since 1955, but this is in part due to inclusion of the American Red Cross which accounts for more than one million dollars of the increase. Goals are set so they can be reached with some margin to spare. The self-fulfilling prophecy of defeat seemed to be operating; failure to make the goal the previous year portends failure for each succeeding year. After making the goals the first year, the community fund was largely unsuccessful until the UF was organized. After 1922 there were only seven victories in the next thirty-two years. The 1923 goal was 27.8 percent greater than the year before, but only 96.2 percent of the goal was obtained. They were not victorious again until 1939. Influentials (individuals and agencies) shirk responsibilities doomed to failure. If the influentials are unable to make the goal, agency executives and board members become dissatisfied. As the head of the UF said, "If one of the campaign leaders fails, he is likely to have a number of good reasons why he cannot help the following year." There are, of course, those seeking status who would try, but the influential who had already achieved status is not looking for ways to have his escutcheon dirtied. So if lower echelon people are recruited for the fund raising job, they have less personal influence on prospective givers of higher socioeconomic status. Consequently, a chief executive officer of a big corporation, if approached by a lower ranking officer of another company to assume a campaign post, will not likely put himself in a position of risk, but asks a vice-president or a manager in his company to work in the campaign. Hence, if the chief executive officers of the largest corporations do not command the fund drive, it is unlikely that they will be brought into lower levels of the campaign organization.

Table 14 gives the amount raised by campaign source for the past thirteen campaigns. This table, in addition to Figure 3 which follows, shows that since the United Fund was organized executive and employee giving has increased as a percent of the total amount raised, but corporate giving has declined. This reflects the unified application of the payroll de-

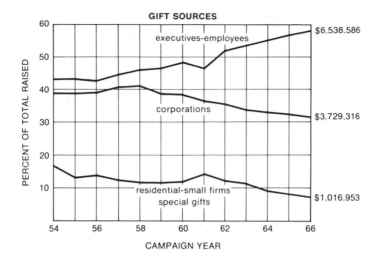

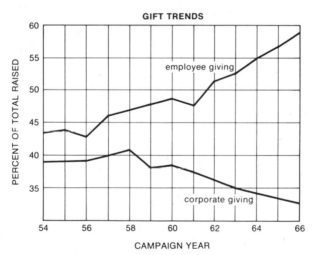

Fig. 3. Giving Sources and Trends—United Fund of Greater St. Louis. (Information for Members of the Board of Directors and Executive Committee of the United Fund of Greater St. Louis, Inc., 1967.)

Table 14

UNITED FUND OF GREATER ST. LOUIS, INC.
AMOUNT RAISED BY CAMPAIGN SOURCE PAST THIRTEEN CAMPAIGNS

Campaign Year	Executives and Employees	%	Corporations	%	Residential & Small Firms	%	Special Individuals & Foundations	%	Miscellaneous	%	Special Anonymous Gifts	%	Grand Totals	%
Community Chest														
1954–1955	$2,535,262	43.7	$2,279,996	39.3	$411,908	7.1	$574,350	9.9	$2,444	.0	$ —		$5,803,960	100
United Fund														
1955–1956	3,608,660	43.8	3,239,920	39.3	544,170	6.6	527,680	6.4	101,224	1.2	223,346	2.7	8,245,000	100
1956–1957	3,571,968	43.3	3,249,500	39.4	552,750	6.7	569,250	6.9	68,742	.8	237,790	2.9	8,250,000	100
1957–1958	3,745,899	45.9	3,272,561	40.1	563,109	6.9	440,694	5.4	26,783	.3	111,954	1.4	8,161,000	100
1958–1959	3,766,825	46.7	3,254,955	40.4	562,000	7.0	410,635	5.1	66,671	.8	—		8,061,086	100
1959–1960	4,057,333	47.2	3,348,776	38.9	602,729	7.0	413,740	4.8	50,234	.6	127,188	1.5	8,600,000	100
1960–1961	4,147,342	48.2	3,357,327	39.0	558,313	6.5	394,903	4.6	84,803	1.0	57,312	.7	8,600,000	100
1961–1962	4,179,739	47.8	3,264,808	37.4	567,597	6.5	643,906	7.4	77,597	.9	—		8,733,647	100
1962–1963	4,893,345	51.6	3,482,226	36.7	508,656	5.4	585,689	6.2	8,032	.1	—		9,477,948	100
1963–1964	5,140,982	52.8	3,470,999	35.6	528,186	5.4	587,343	6.1	13,010	.1	—		9,740,520	100
1964–1965	5,587,040	55.1	3,528,791	34.8	470,096	4.6	531,376	5.2	35,660	.3	—		10,152,963	100
1965–1966	6,140,636	56.6	3,638,460	33.5	476,476	4.4	563,218	5.2	33,132	.3	—		10,851,922	100
1966–1967	6,538,586	57.7	3,729,316	32.9	461,685	4.1	555,268	4.9	43,692	.4	—		11,328,547	100

Information for Members of the Board of Directors and Executive Committee of the United Fund of Greater St. Louis, Inc., 1967.

duction plan in the larger corporations. Also, it has apparently enabled the corporations to keep their own giving about the same. This means a broader base of employee giving has been achieved as employees and executives of corporations are contributing proportionately more. However, the residential and small business giving which are combined in the tabulation has decreased absolutely and relatively in terms of percentage of total amount raised. Also, even though there is an absolute increase in corporate giving, its proportion declined from 39.3 percent to 32.9 percent. The foregoing suggests that corporate leadership has more control over their own executives and employees than over small business. Obtaining a greater share from their employees enables the corporate philanthropic dollar to be distributed more for other types of welfare and for educational purposes. Today, goals are set in terms of what the men from Civic Progress think can be taken from the community. Agencies do have an opportunity to indicate their needs but the "need gap" is usually several million dollars and shrugged off as being unobtainable or too risky to try to get. When David R. Calhoun, president and board chairman of St. Louis Union Trust Company and chairman of Civic Progress, was asked (in an interview with the author, March, 1967) about the three percent increase in the UF goal per year between 1960 and 1965, he said: "It isn't enough, is it?" In explanation, he said that some of the agencies are so expensive to operate and exemplified this with the per capita costs of day care centers. He indicated a need for more management analysis and cost effectiveness studies in agencies. He added that it was also the matter of making the goals: "Sometimes we set them too low, I realize. But it is so important for the people to make the goals, and for the image of the community."

The involvement of Civic Progress is shown in the following. W. Ashley Gray, Jr., President of General Steel Industries and a Civic Progress member, was asked by a small group of executives speaking for the UF to be campaign chairman for the 1966–67 drive. Mr. Gray was quoted as saying: "When I was asked, last January, to be campaign chairman, I was very flattered. Some of the city's top leaders were around the table. But my feeling was one of misgiving; it was a mountain I didn't know if I was going to be able to climb" (Wood, 1966). According to the *Globe* reporter, he turned to James P. Hickok, board chairman of First National Bank, one of the men at the table, and said that he couldn't do the job unless all present guaranteed they would help. "Ash, I will take any job you assign me," Mr. Hickok replied—and Mr. Gray promptly assigned him special gifts division chairman. Going around the table, he found other willing recruits for top jobs in the 1966 campaign. Many of the men were former chairmen themselves. In fact, Gray noted, eleven of the men who held important jobs in this campaign had been campaign chairmen or

president of the United Fund in past years (Wood, 1966). A number of fellow Civic Progress members besides Hickok rolled up their sleeves to help. There was Frederic Peirce, UF Board President this year; M. R. Chambers, head of Pace Setter Gifts; Harry F. Harrington, head of the Banks Division; Raymond E. Rowland, Chairman of the National Corporations Division; J. W. McAfee, Sales Training Committee, and Ben F. Jackson, Accounting. Gray praised the work of the Public Relations Committee, one of whose members was August A. Busch, Jr. David R. Calhoun, Jr. was singled out for recognition for his job as head of the Loaned Executive Committee. Calhoun contacted more than 100 corporation presidents and was able to get them to lend more executives than in any previous campaign. Loaned executives serve under the professional staff in the UF, and solicit companies. This was one of the innovations of the United Fund.

The most important social devices for the application of power are the control over in-plant solicitation and payroll deduction. The big health agencies—heart, cancer and polio foundation are simply excluded. The logic back of solicitation in the company limited to the Fund is that it is supposed to be one gift for all. Giving is, of course, more liberal when it can be spread out over a period of time through payroll deductions. The latter is becoming increasingly automatic. A plan is now provided for the employee to sign a card agreeing to an automatic pay check deduction defined by his corporation and the UF as a fair share.

The UF publicly denies compulsion. In the handbook used in the 1966–67 campaign, one of the commonly asked questions is answered:

> I hear that some companies force their employees to give. How about that? The United Fund believes that every employee should receive an adequate explanation concerning the importance of United Fund agencies' services, and that questions should be answered. A proper approach by a volunteer solicitor includes an adequate explanation of the services. However, under no circumstances should an employee be threatened or coerced into contributing. Giving is and must be a voluntary act (*United Fund Handbook*, 1966–67 Edition: 9).

In the interview with Calhoun, he said: "You have to twist a few arms." He mentioned talking to some girls in his office who live in Illinois about giving. (The UF does not finance agencies on the east side although many people from Illinois do work in St. Louis.) There is the implicit threat of what the employee imagines might happen: it could mean not getting a raise or not being promoted. The individual who does not give or gives less than is expected violates a norm set by both management and labor.

The United Fund is organized like a business corporation. It has an executive vice-president with functions similar to those in business. The remoteness or virtual autonomy of the executive committee from the board of directors and even more distance between the executive committee and the agencies is analogous to the business corporation which operates rather independently of the board of directors, the stockholders, and the general public. This oligarchy of the money-getters is congruent with Michels' theory (1959:32) of oligarchical tendencies of organization, because under the UF there is rule by a small "minority of directors" over a "majority of directed." This has been intensified since the end of the Community Chest era. The stratification of organizations is apparent in the involvement of the topmost corporate leadership at the highest levels of the campaign. This maximizes the opportunities for reciprocities and, of course, for sanctions for corporations or individuals within the corporations who refuse to take part in the formalized and depersonalized gift-giving transactions. Finally, there is a noncontroversial area of community service which can legitimate the notables as civic leaders, and compensate for other activities which are more transparently self serving such as the Civic Center Redevelopment Corporation. Policy is therefore controlled by those who hold the corporate wealth. The implementation of policy depends upon money and manpower. Civic Progress controls both in the corporations and the private agencies responsible for policy setting.

Returning to the Council take-over, the following (*1911/1961 Fifty Years of Community Service,* Annual Report, Health & Welfare Council of Metropolitan St. Louis) briefly stated the new direction in 1961: "strong emphasis on laymen's responsibility backed up by a professional staff. . . . In recognition of the changing strengths of agencies and needs of the community, the work of HWC was increasingly geared toward community centered problem solving with emphasis on planning by fields of service." The major issue was over control of social policy. The large companies wanted to have more say-so which meant that agencies in the UF and Council would have less. Pressure for this kind of change forced the resignation of the Council's executive director in the mid-fifties, and the executive who followed was not able to meet UF demands and left in about a year. The third man, who followed, was able for nearly four years to provide an uneasy and conflictual stewardship for the Council.

Civic Progress and its corporate executive functionaries expected the Council to orient itself to the needs of the UF, consult with community leadership, work on projects, and deemphasize being a broker fostering agency services. About this time, the UF executive vice-president and the president of the board of directors of the UF were appointed ex officio members of the Council's board of directors. This has continued.

In 1960 the Council was asked by the UF to do a study of the social welfare resources and needs of the metropolitan area (Figure 4) and to make recommendations for bringing resources in line with needs. Appar-

ST. LOUIS
STANDARD METROPOLITAN AREA
1960

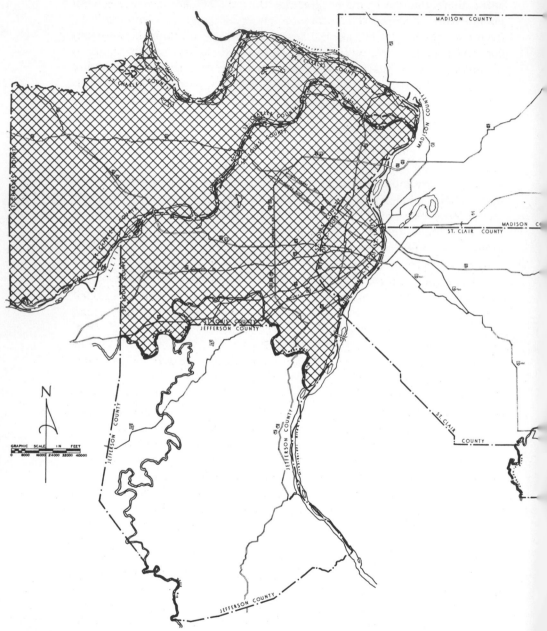

RESOURCES-NEEDS PROJECT STUDY AREA

Fig. 4. Resources-Needs Project Study Area.

ently, the basic purpose, from the standpoint of the UF, was to indicate what agencies should be cut, left the same, increased, dropped or merged. Instead of dealing with such pragmatic matters, the Resource-Needs Project spent much time creating a lexicon of service classifications and elaborate definitions, a crude systems analysis of community problems, and laboring over ranking the seventy-nine community services in terms of gravity of condition to which the service is directed, effectiveness of service, current quantity, current quality, and expected change in requirement (in five years) (*Seventy-Nine Services, A Supplement to Guide for Community Action, The Report of the Resources-Needs Project,* Health and Welfare Council of Metropolitan St. Louis, February, 1963). Although the study represented an excellent exercise in the formulation of social policy, it was too general to satisfy specific, fact-minded Civic Progress influentials. The *Guide for Community Action* published in 1963, three years after the original request, did not recommend action concerning any particular agency. The planning directors of the Council studiously avoided such particularity. Agencies were listed under fields of service such as Family and Individual Counseling, Family Life Education, Financial Assistance, Group Services-Socialization; it was not possible to infer which ones should get more or less. The recommendations for increased funding applied to a set of agencies whether individual agencies might need or not need additional funding. Thus, the voluminous volumes were not guides to action, according to some leaders, but rather to further study.

Some agencies dear to Civic Progress did not receive the highest rankings. Scouting and the "Ys" classified under Group Services-Socialization (along with a number of other agencies) although given the highest rating for Gravity (consequence) did not receive the highest for Pertinence. Instead of receiving: "Is known to be pertinent" this field of service got "generally recognized to be pertinent." Likewise on effectiveness, it was awarded "relatively high" and not "high." Concerning quality of service, it was rated as "needing much improvement." This is the lowest quality rating used on the scale. Scouting and the "Ys" are agencies with which business influentials have been closely identified, and some were reportedly displeased with the ranking. Being ranked as a field of service, along with other agencies meeting the definition of having socialization activities, they suffered the ignominy of being rated the same as agencies which were seemingly of lower caliber. Included in this field of service, with the favored agencies, were ones which had been less blessed through the years: namely, those agencies which attempted to serve the black, the poor, and the disadvantaged: for example, settlement or neighborhood houses. To the businessman, this included dissimilar agencies and arrived at erroneous conclusions concerning needs. The quality of services for underclasses was admittedly inferior.

The Council was accused of wasting thousands of dollars for something useless for the Fund, meaningless to the community, and unresponsive to everyday needs of agencies. As pointed up in a previous chapter, this was a time when the voluntary agency's role was seriously challenged. The staff of the UF, as well as certain business influentials, contended that the Council was not in touch with the real world of social agencies and substituted abstractions for practical matters. Recommendations, in the *Guide for Community Action,* to spend millions of private and public dollars were too vague. It placed Civic Progress leadership in the awkward public position of having inadequately met organizational and financial needs of health, welfare and recreation services.

In 1962, during the time the Resource-Needs Project was conducted, Civic Progress took hospital planning out of the Council and set up the Hospital Planning Commission. The origin of HPC is indicative of Civic Progress' sensitivity to the need to exert control where corporate influence is declining because of third party payments, and hospital depreciation reserves being accumulated. It is also important as an instance of Civic Progress' control over HWC. Under the Health and Welfare Council, the hospital division was composed of delegate members from the health and hospital field. It was a council within the Council concerned with services and facilities. Civic Progress set up HPC without delegate representation but with HPC's board interpenetrated with Civic Progress members. Funding came from the City, UF, and a research grant from the United States Public Health Service. This of course left only health planning in HWC.

In February of 1963 the following was reported in the Council's *Newsletter* (The Health and Welfare Council of Metropolitan St. Louis, February, 1963):

> Negotiations have been underway in the last couple of months between Health and Welfare Council and United Fund in regard to trend of program for the Health and Welfare Council. It has been considered that the Health and Welfare Council is "spread too thin" and is engaged in too many broadside surveys and studies—not adhering to the kinds of agency studies and requests that United Fund needs for its own concerns. Negotiations resulted in complete autonomy and present budget for the Health and Welfare Council, but with the understanding that several kinds of Fund agency studies need to be undertaken and completed by the Health and Welfare Council.

COUNCIL ENSLAVEMENT

With the publication of the *Guide for Community Action* in 1963 two planning directors and the research director resigned. Relations between Council and Fund further deteriorated. In December the Council received an ultimatum in the following letter:

UNITED FUND OF
GREATER ST. LOUIS, INC.

St. Louis, Missouri 63101
December 29, 1963

Mr. John H. Poelker, President
Health and Welfare Council of Metropolitan St. Louis
1006 Market Street, Room 212
St. Louis, Missouri 63103

Dear Mr. Poelker:

The United Fund allocation to the Health and Welfare Council for the 1964 calendar year has been approved at $189,739, subject to certain stipulations. In the interest of clarity concerning these, the entirety of the United Fund Executive Committee resolution was as follows:

> Be it resolved, and it hereby is, that the Executive Committee approves for the calendar year 1964 the same allocation to Health and Welfare Council as that made for 1963, $189,739, and,
> Be it further resolved, that this allocation be subject to the provisos set forth in the Agencies Relations Committee report, and,
> Be it further resolved, that a committee of the United Fund shall be appointed to meet jointly with a similar committee of the Health and Welfare Council to evaluate the services of the Health and Welfare Council, and,
> Be it further resolved, that the United Fund committee so appointed shall report quarterly to the Executive Committee concerning whether or not the spirit of the arrangement, including the Maxwell Committee report, is being carried out, and,
> Be it further resolved, that payments to the Health and Welfare Council against its 1964 allocation of $189,739 shall be likewise reviewed quarterly in terms of the Health and Welfare Council's compliance with the arrangement.

The provisos specified in the second paragraph of the resolution, as contained in the Agencies Relations Committee report, are:

(1) The United Fund is to prepare specific procedure for agency program evaluation by members of the Council staff.
(2) A list of agencies to be so evaluated during 1964 is to be prepared with quarterly deadlines established for completion of specific agency reports.

(3) The Council is to allocate staff time and related overheads approximating $100,000 for these agency evaluations.

(4) The Council is to allocate approximately $50,000 to long-range planning or "the big picture."

(5) The Council is to allocate the remainder or approximately $39,739 to certain "tie-in" services. These would include the Information and Referral Service and certain other activities, primarily of a research nature. The latter would include United Fund agency service statistics and projects specifically requested by the United Fund such as the Beneficiary Study.

The first two of these have been fulfilled by the United Fund in the memorandum, "Review of United Fund Agencies by the Health and Welfare Council," dated November 26, 1963.

As you know, the United Fund Annual Meeting will be held shortly, at which time a new president will be elected. It will be his responsibility to appoint the United Fund committee authorized in the Executive Committee resolution.

Cordially,

(*Signature*)

Frederic M. Peirce, Chairman
Agencies Relations, Committee

FMP:p

cc—Mr. Howard B. Hollenbeck

Poelker is Comptroller of the City of St. Louis and Peirce is a member of Civic Progress.

In February 1964, following the take-over of the Council's budget, UF executive committee asked justification for keeping health planning in the Health and Welfare Council. In defense of leaving health planning in the Council the following was prepared by the planning director for health services (Memorandum of Health & Welfare Council of Metropolitan St. Louis, Statement on Council Responsibilities in the Health Field, Feb. 25, 1964):

Question has been raised as to why the Health and Welfare Council continues to require the services of two professional staff persons (actually one full time person and one three-fifths time) in view of the existence of the Metropolitan St. Louis Hospital Planning Commission.

When the Hospital Planning Commission was established it was with the understanding that this commission would concentrate on problems relating to capital fund campaigns, expansion of hospital facilities, increase or decrease and use of numbers of hospital beds in the community, and other matters relating to the capital expansion of facilities. It was also assumed that the Health and Welfare Council would continue to work with projects and aspects of the total health field—relating to hospitals as these programs had bearing on hospital facilities, out-patient services, etc.

It is within that framework that the Health and Welfare Council and the Hospital Planning Commission have worked very closely together in the last couple of years, collaborating in the issuance of reports and sharing of material, etc.

In addition it was stated in the same Memorandum that the purpose of health planning was:

To enable the community, represented in the Board of Directors of the Council, to identify health problems and to deal with them efficiently through appropriate direct service organizations.

To provide consultation to help agencies in aiding them in their planning and to help assure that their services are effectively administered.

The budgetary controls and the attempted take-over of health planning by HPC were irregular developments in community planning. I know of no other case where a health and welfare council was literally put in the "hands of the receivers." The "receivers" in this particular case was John R. Bartlett, Jr., Partner, in Peat, Marwick, Mitchell & Co. Bartlett, immediate past President of HWC, was appointed chairman of the Agency Review Committee. His responsibility was specified in the Peirce letter to Poelker.

The agency review was to give UF information which its staff and some civic elites considered necessary for specific decisions about agency effectiveness and funding. The UF assigned most of its agencies to be reviewed in a year's time, and the Council's planning directors developed an outline, approved by Bartlett and the UF staff, to be used for the reviews. Tension between Council and UF staff was high, the former expressing much hostility about the repressive and unreasonable nature of the demands. Planning directors contended that the job was too much to be done with the available staff.

An agency review involved a planning director or his associate (only one of the three planning directors had an associate) spending specified lengths of time with each phase of the study. The time then was to be programmed and accounted for to John Bartlett. This required not only the cooperation of the council staff but the staff of the agency being studied.

Also the executive director of the Council was supposed to supervise the planning directors to see that they carried out the reviews and did not involve themselves in other activities which would prevent them from staying on schedule. Since not all the Council time was tied up in agency reviews, it was difficult to know when something should be given priority over an agency study. Bartlett was particularly interested in obtaining high quality reviews because he wanted to impress UF business influentials. He had a pretty good idea of what should be done because his firm made management studies.

Bartlett set up regular appointments with the Council staff to determine what stages the different reviews were in, and how many hours had been used for each phase. He read and edited reviews; some reviews were returned because they did not measure up to his standards. Lack of specificity was commonly cited. In some instances, planning directors went back to the agencies for more information. Consequently, kinks developed in what was otherwise rather tight management control.

To further complicate the matter, the planning director for group work and recreation services resigned, and a few months later the planning director for health services joined the Hospital Planning Commission as associate director. So even with rigorous auditing procedures to account for Council expediture of its budget, production of agency reviews simply got out of hand.

Adding more to Bartlett's consternation was that the Council's executive director became more resistive and defensive as Bartlett's distress mounted over the inability of the Council staff to complete the work as required.

Also, St. Louis' War on Poverty created pressures on the Council to respond. Agency reviews stood in the way of what planning directors considered to be far more significant involvement.

Bartlett assigned a junior executive from his company to HWC's staff to work on reviews. The purpose was apparently to cope with a lagging output and to assess what might be done to speed up performance. Bartlett's executive found it difficult because of his lack of social work expertise. He spent much time in consultation with planning directors. Moreover, his entry into the review process was viewed with alarm, because it is one thing to be studied by a social work insider and quite another by a management consultant. Perhaps other Council functions, some thought, could be taken over by management experts. The upshot of it all, was that Bartlett's man had trouble, too, in getting things done and was eventually transferred back to the company.

In Summer 1964, Poelker, the Council's President, asked for the Council executive's resignation. A planning director then assumed the role of coordinating staff activities on reviews with Mr. Bartlett.

COUNCIL REGAINS POWER

The board of HWC quickly recruited another executive director. Dan MacDonald (degrees in social work and journalism) who had been the director of United Community Services in Omaha, Nebraska was employed. Omaha has a combined fund and council; thus, MacDonald presumably had the kind of background which could go well with a council forced to link itself closely with the UF. A fund raiser would likely share the business ethos required to assure more harmonious and effective relations between UF and HWC. He joined the staff early in 1965. He is a Catholic like Poelker who was President of the Council's Board of Directors. Duncan Bauman, Poelker's successor to the presidency, then General Manager for the *Globe-Democrat* and now its Publisher, is also Catholic.

Relations with the UF immediately became more cordial. The new director referred to it as the "honeymoon" period. The Hospital Planning Commission which had been set up by Civic Progress was returned to HWC. However, instead of being restored as a function of the health planning director, it remained a separate agency while sharing office space with the Council. MacDonald became the Hospital Planning Council's head in addition to being Executive Director of HWC. HPC retained its own board sprinkled with Civic Progress members.

The agency reviews went on but at a much reduced tempo. The attitude changed; agency reviews had lower priority. MacDonald learned that the Fund had not used reviews effectively because some recommendations could lead to increasing allocations to agencies. Although some agency programs were improved and the UF made some cuts, the net results were less than satisfactory. So it was decided that studies could realize more if worked out on a selected basis. Doing fewer agency reviews but in greater depth as needed could be more fruitful. Broad-scale study of many Fund agencies yielded superficial facts and less useful conclusions than the Fund had originally thought. Hence, MacDonald got the Council relieved of the reviews and John Bartlett received a plaque for his stewardship.

Likewise, the beneficiary study, which the UF insisted that Council do, was not used effectively. UF agencies kept records on numbers of recipients served, by place of employment and by area of residence. Two hundred employers were listed by the UF. Agencies made tallies when they had contacts with employees from firms on the list. The recipient's place of residence was coded on another form by postal zone for the city and municipality for the county. As it turned out, the Fund made no significant use of the beneficiary data in campaign publicity. For some employers, whose employees did not get much service from UF agencies, the beneficiary study would not provide a persuasive argument for giving. They were reluctant to give much publicity to area figures because underserved municipalities could use these facts against the UF. Such arguments against doing the beneficiary study in the first place had been presented by the Council staff, but to no avail.

John Dillencourt, Executive Vice-President of the UF and two agency relations directors resigned prior to MacDonald's coming. Apparently business influentials were less than satisfied with the UF staff.

The fifty-fifth year of the Council never looked brighter, and its annual report never more posh. The mark of corporate elites is reflected in design and layout. Fred Peirce, CEO of General American Life Insurance Co., was on the board that year and the Council strode forth to meet the deepening urban crises.

The measure of a social agency is its executive director. The better the agency, the better the executive and vice versa. In Dan MacDonald, civic influentials found a leader whose stature was commensurate with the kind of agency they saw needed. One can surmise that when Civic Progress did not expect much from the Council, it allowed executive incompetence in the Council to drag on. From an organizational point of view this can be functional in undermining confidence in the agency and creating a social ground receptive to oligarchical decision making. Of course, this principle applies to public as well as to private polities. That is, governmental impotence results in strong private organizations assuming responsibility for a mass society.

In closing this chapter, quotations from Dan MacDonald's speech in early 1967 ("Urban Needs—The Search for Answers," Citizen's Conference on Community Planning, Oklahoma City, Feb. 2) assert his allegiance to the private system and its power to command public institutions:

> There are several groups of people here for whom this conference should have meaning. First, this is a "Citizens Conference." Its direction, its purpose is based on the principle that the citizen should lead and be responsible for community planning wherever he lives. [*It should be noted that he spoke to affluent civic influentials.*]
>
> A second group is the practitioner, the "pro" from local communities who is responsible for program in the health and welfare planning council or united fund or perhaps both operations, and who most likely is and certainly should be up to his neck in community affairs. [*He referred to organizations whose cadre are brokers for these affluent corporate influentials; they are not catalysts for the disadvantaged.*]
>
> A third group is a special one—representative of governmental agencies. They have come to learn about councils, if they don't know them already; to help us chart the course of cooperation with government; to discover if councils have the determination and courage to

help our cities or are content to remain with the comfortable issues and roles of the past.

The Conference theme centers on responsibility. *Whose Responsibility is urban planning?* This, as far as I am concerned, is also one of our major urban problems. Planning efforts and planning mechanisms have proliferated at a phenomenal rate. Planning at the urban level is as complex as any piece of machinery we have divined. We have innumerable local, overlapping districts for health, public safety, welfare, education and economic development. My community has 95 separate municipalities, 174 taxing authorities and 420 taxing areas in just one county. We have state agencies, new state and regional planning agencies. Very few people at any local governmental planning level, let alone in voluntary planning organizations, have mastered the risky art of dealing with federal agencies. . . .

In my community there are at least fifteen to twenty organizations with large areas of planning. All of these organizations are independent of each other and operate closely together only if relationships have been established by their respective executives and/or boards or by some marriage binding each other together, such as in the use of money. This, of course, is a most effective inducement for getting along together. [*Of course, he glosses over the thoroughgoing interlocking and interpenetration of strategic agencies not only in funding but in officerships. Some planning organizations do not do much anyway, but they do serve the pluralist myth.*]

There is another dimension. Nearly all major federal legislation in the past two years has made some reference to comprehensive planning, or the relation of specific programs to other aspects of community life. The regional medical program refers to "cooperative arrangements among public or private institutions or agencies." The Model Cities Act calls for "the most effective and economical concentration and coordination of . . . public and private activities. . . ." Public Law 89-749 is revolutionary in relation to health. It states that "comprehensive planning for health services, health manpower and health facilities is essential at every level of government." Its purpose is "to assure comprehensive health services of high quality for every person." [*The foregoing show the elitist character of community planning. Comprehensive planning often leads to more totalitarianism and more bureaucracy. Note well the idea of little people having anything to do with this process is not uttered by MacDonald. It would be inappropriate for him to say as executive director of a council.*]

[*A rejection of Galbraith's technostructure is evident in his saying:*]

Last September, *The Wall Street Journal* quoted William I. Spencer, as follows: "What we are confronted with is virtually a new language of stocastic processes, minimax, algorithms, heuristics, parameters, models and simulation, all of which are somehow related

to systems analysis. This is simply another name for operations research. . . . But unfortunately, operations research is becoming a
fad. In the process, the admission is beginning to leak out that, regardless of the precision achieved by the new aids and tools, policy
decisions will rest fundamentally on institutions and human judgment."
[*Of course, the crucial questions are what institutions? Whose judgment?*]

Now, what about health and welfare councils? As to the councils'
relation to government: Close to seven out of eight community health
and welfare dollars are expended through some governmental agency.
The council cannot plan only with the voluntary dollar or the voluntary
agency but must be concerned with total expenditures. One must be
related to the other. Our largest problems, e.g., public welfare, have
the greatest need for our services because of their limited planning
ability and little citizen interest. I would estimate that the "new"
council spends somewhere near two-thirds of its time directly or indirectly in the governmental agency arena. [*He includes public money
not included in the periodic study of health and welfare expenditures
conducted under the auspices of United Community Funds and Councils. Who are the citizens to do this planning? Elites. Who are the
cadre to do this planning? Elites.*]

What are the councils' planning assets? Unquestionably, the most
valuable is its citizen base. Councils have developed citizen participation to a greater degree than most governmental agencies or perhaps
any other field with the possible exception of education. There is a
breath of fresh air about a board, a group of people, an organization
which can express itself critically and constructively in the community interests.

[*He advocates the interpenetration of government by these civic
influentials.*] The voluntary element or citizen participation in new
governmental programs and in urban planning must be developed
as never before. Somehow or other we in the United Way movement
have told ourselves for years that we are one of the last bastions of
voluntarism. I would suggest that we not delude ourselves into
thinking that the volunteer wants a role of preserving insignificant
or outmoded services or sees the voluntary agency as his only avenue
for civic service. Voluntarism shifts over a period of time and it
would be a mistake to think that it would not leave federation if
there was a better way to solve problems or meet the needs of people.
I would hope that we would lead the way in placement of council and
United Fund volunteers in all programs of a public nature.

[*His perceptiveness concerning the role of industry in welfare
programs for profit is indicated in the following.*]

Along the lines of innovation I would suggest close examination
of the trend of business and industrial involvement in the development
of programs in the human resources field. This is a relatively new,
untapped area. Job Corps, health systems, social problem management

are becoming industry specialists. I believe that business has a capacity and talent in human welfare that needs local exploration and exploitation.

What we have here is *no* failure to communicate. The stage is set for elites; the actors are elites and the play is elitist. It is a top down type of planning with a social work cadre serving as brokers for business influentials and agencies who manage the poor.

UF-HWC-HPC is one of many patterns that make it seem that St. Louis is pluralist but as much as we might wish it otherwise, the jigsaw pieces fit. The overall pattern is concentration of economic power guiding a seemingly disconnected but actually interconnected process. Civic Progress gives the field magnetic attraction for bringing the pieces together.

CONCLUSIONS

The transition from a more democratic to a less democratic model of urban power and social welfare is traced. The take-over of the Community Chest by Civic Progress is documented. The increased efficiency of the new agency, the United Fund which replaced the Community Chest is evident. The doughty performances of civic notables of the highest rank is apparent. In addition to their names, they lend substantial amounts of time and their honor to battle for community funds. These civic noblemen may be likened to knights of the roundtable. Their power is exercised victoriously in public ways.

In being guardians of public trust, they are ruthless when deemed necessary. Their subjugation of agency kingdoms such as the Health and Welfare Council testifies to their Machiavellian sagacities. None but they can, with such effectiveness, push other agencies around as pieces on the community's chess board. They are bestowers as well as executors. And they are pragmatists par excellence.

Thus the manipulation of the United Fund, Health and Welfare Council, and the Hospital Planning Commission by Civic Progress is illustrative of the United Way. The model is private ideology as civic elites go about shaping social welfare as they like it.

Chapter 7

Sectarian Organization

of Social Welfare

Since a considerable amount of health and welfare services are under Protestant, Catholic, or Jewish auspices, the central question in this chapter is what groups control these services? How autonomous are activities under religious auspices? Sectarian auspices refer to sponsorship. However, the sources of power may be either clerical or nonclerical in nature. A parallel question to be dealt with here is how control over sectarian welfare has been affected by increased use of public funds.

JEWISH ORGANIZATION OF SOCIAL WELFARE AGENCIES

Both the Jewish Federation and United Fund plan and budget for Jewish agencies in the City of St. Louis and St. Louis County. These agencies are for an estimated 57,500 Jews of whom less than 1 percent reside in the city.[6] Most Jewish agencies such as Jewish Hospital and the Jewish Employment and Vocational Service also have non-Jewish clientele. Prior to World War II, most Jews lived in the city, but as their economic status improved, combined with the in-migration of Negroes, the Jewish population was radically redistributed.

The Jewish Federation, organized in 1900, raises money, plans, and budgets for agencies in the city and county. In addition to campaigning for local Jewish agencies, the Federation secures contributions for state, national, and overseas social welfare programs. The Federation is a member agency of the United Fund and receives part of its own operating budget from this source. Although originally independent of the UF, because of the economic effects of the depression during the 1930s, the Federation became aligned with the Community Fund, a predecessor agency of the United Fund (Rosen, 1939:108–111). In 1932 the Community Fund, which at that time was a federation of Protestant and nonsectarian agencies, asked for a joint fund drive of the Jewish and Catholic Federations. This federation of federations was named the United Relief Drive. Each federation was to retain autonomy in the budgeting of their member agencies. The agreement also stipulated that the Jewish Federation would continue to raise funds for all Jewish agencies not included in the United Relief Drive.

Although both Catholic Charities and the Jewish Federation presently participate in the budgeting of United Fund money, distribution of the funds does not depend upon formally constituted committees which represent Catholic or Jewish interests per se. In short, determination of allocations has been secularized.

Of the $10 million received by the selected Jewish agencies shown in Table 15, all but $100,000 went to agencies belonging to the United Fund.[7] The agencies not belonging to the UF obtained more than $800,000 from the United Fund. Table 15 also shows the year of agency organization, giving a measure of new agency formation. The foregoing suggests that the United Fund is influential in the Jewish social welfare field.

The Federation's board is given in Table 16. Most Federation directors are entrepreneurs; few of its members are executives of the large locally based corporations. Although Jews in Civic Progress are in the United Fund, these same Jews are not on the board of the Jewish Federation. Despite the absence of topmost leadership from the board of the Federation, Jewish Hospital and the Jewish Community Centers Association do have this upper stratum of business elites. The inference is that the Federation lacks the community wide appeal enjoyed by Jewish Hospital or the Jewish Community Centers Association, and hence the status associated with prestigious agencies. Nevertheless, the Federation board members cluster in the wealthy suburbs of Ladue and Clayton.

The United Fund, Arts Council, and the Symphony Orchestra are examples of drives which enlist influential Jewish leadership (Interview by author with William Kahn, January 3, 1967, St. Louis, Mo.). However, in return for their efforts, Civic Progress influentials assisted the Jews in building the new Jewish community center.

Also there is a split between German and Russian Jews in St. Louis, with the former being more affluent but with ambivalent attitudes about the State of Israel, and consequently not as responsive to Federation needs.

Table 15

JEWISH AGENCIES BY MAJOR SOURCE OF FUNDS,
TOTAL INCOME AND YEAR ORGANIZED

Agency	Major Source of Funds	Income in 1965	Year Organized or Incorporated
Jewish Hospital	Fees	$ 7,105,700	1900
Jewish Community Centers Association	Fees	1,209,186	1952
Jewish Center for Aged	Fees	848,296	1906
Jewish Employment and Vocational Service	Public	345,966	1940
Jewish Family and Child Service	Contrib.	246,603	1963
Jewish Federation of St. Louis	Contrib.	92,901	1900
Council of Jewish Women	Contrib.	65,271	1917
Jewish Community Center Day Nursery	Contrib.	43,450	—
Jewish Community Relations Council	Contrib.	40,229	1938
Anti-Defamation League of B'nai B'rith	Contrib.	15,940	—
B'nai B'rith Hillel Foundation	Contrib.	14,290	1946
United Orthodox Jewish Committee	Contrib.	8,911	1923
American Jewish Committee	Contrib.	8,509	—
Noshim Rachmonioth Society	Contrib.	8,484	1907
Total		$10,053,736	

Organizations for which income information was unavailable: Ben A'Kiba Aid Society; B'nai B'rith Youth Organization; Jewish Foundation for Retarded Children; Jewish Shelter & Aid Society; Jewish Special Needs Society; Miriam School; Organization for Rehabilitation Training, St. Louis Region; Scholarship Foundation of St. Louis, and the Life Seekers.

Concerning use of public funds, the Jewish Employment and Vocational Service points up certain implications for private control. The former director (Bernstein, 1965:6), in a paper concerning public funds and the private agency, commented on a tendency for the private funding body to reduce its allocation as the private agency received more public funds. Being subsidized by public funds and having many non-

Table 16

BUSINESS AND RESIDENCE OF BOARD MEMBERS
OF THE JEWISH FEDERATION, 1967

Name	Business	Residence	Postal Zone
Abramson, E. D.	Central State Paper & Bag	Creve Coeur	41
Bierman, Arthur	Central Waste Material	Olivette	32
Bettman, Irvin, Jr.	Marx-Haas	Ladue	·24
Cohen, Stanley	Central Hardware	U. City	30
Cutter, Jack R.	Cutter Karcher Shoes	U. City	24
Dubinsky, Melvin	Jack Dubinsky & Sons Realty	Ladue	24
Dubinsky, Saul	Dubinsky Realty	Ladue	24
Edison, I. J.	Edison Bros. Stores	Ladue	24
Einstein, Major	First National Bank	St. Louis	08
Fixman, Ben	Fisher-Fixman (Diversified Metals)	St. Louis	41
Fleishman, Alfred	Fleishman-Hilliard Pub. Relations	Ladue	24
Frank, Harris	American Recreation	Creve Coeur	41
Gallop, D. P.	Attorney	Ladue	24
Goldstein, I. E.	Wiles-Chipman Lumber Co. Inc.	Ladue	24
Gross, Lester	Gross Engineering	U. City	32
Hausfater, Mrs. R.	Roberts Record Distributor	Ladue	24
Hearsh, Howard	Tension Envelope	Ladue	24
Karl, Michael, M.D.	Physician	Pasadena H.	17
Katzenstein, Rabbi M.	Rabbi	Clayton	05
Kaufman, Lee I.	Kaufman & Wise Insurance	Rich. Hts.	17
Krainos, Morris	United Clothing Co.	U. City	30
Krupnick, Sam	Krupnick and Assoc. Pub. Relations	Clayton	05
Levy, Willard	Angelica Uniform	Creve Coeur	41

Jews in its program, JEVS is unique among United Fund agencies who are also members of the Jewish Federation, because only specific JEVS programs receive supplemental funding from the Federation (Interview by author with Harry Kaufer, September 9, 1967, St. Louis, Mo.).

Table 16 (continued)

Name	Business	Residence	Postal Zone
Loeb, Alexander	Renard Linoleum	Clayton	05
Lowy, Fred	Fred Lowy Linoleum	Clayton	05
Marshall, Kenneth A.	A. & S. Aloe Co.	Clayton	05
Marx, Richard	Attorney	Clayton	05
Messing, R., Jr.	World Color Press	Westwood	31
Meyer, Julian	Salomon Bros. & Hutzler	Clayton	05
Millstone, I. E.	Millstone Construction	Brentwood	44
Packman, Victor	Attorney	Clayton	05
Pearlmutter, M.	Edison Brothers	Ladue	24
Price, Elmer	Attorney	Clayton	05
Raskas, Ralph	Raskas Dairy Inc.	U. City	30
Ruwitch, Joseph	Renard Linoleum	Rich. Hts.	17
Salomon, Sidney, Jr.	Sidney Salomon & Assoc.	Frontenac	31
Scherck, Gordon I.	Scherck, Stein & Franc	Clayton	05
Schweich, Edward	Cerro Corporation	Westwood	31
Seidel, Herbert	Laclede Gas	Creve Coeur	41
Seltzer, William	Real Estate	U. City	30
Seltzer, Mrs. W.	Housewife	U. City	30
Shenker, Morris A.	Attorney	Town & Country	31
Shifrin, Robert	Kaufman Wise Inc.	Clayton	05
Soffer, Harry	Attorney	U. City	24
Stein, Elliott	Scherck, Stein & Franc	Ladue	24
Sussman, Earl	Attorney	Frontenac	31
Wolfort, Jesse A.*	Attorney	U. City	30
Wolfson, Robert	Feld Chevrolet	Clayton	05
Yalem, Charles	ITT Aetna Finance	Ladue	24
Yalem, Richard	ITT Aetna Finance	Ladue	24
Zorensky, Louis	Shopping Center Developer	Ladue	24

* Deceased.

Board lists of agencies in addition to Polk's St. Louis City and St. Louis County Directories.

In respect to Jewish Hospital, and other private hospitals as well, the United Fund expects Medicare to enable UF dollars to be withdrawn entirely (Interview by author with David Gee, December 27, 1966, St. Louis, Mo.).

Table 17

CATHOLIC AGENCIES BY MAJOR SOURCE OF FUNDS,
TOTAL INCOME AND YEAR ORGANIZED

Agency	Source of Funds	Income in 1965	Year Organized or Incorporated
St. John's Mercy Hospital	Fees	$ 6,664,293	1890
St. Mary's Hospital	Fees	6,567,636	1924
De Paul Hospital	Fees	5,024,818	1930
Firmin Desloge Hospital	Fees	3,772,670	1933
St. Anthony's Hospital	Fees	3,190,347	1872
St. Joseph's Hospital (Kirkwood)	Fees	2,541,391	1954
Cardinal Glennon Hospital	Fees	2,340,276	1956
Alexian Brothers Hospital	Fees	2,114,240	1839
Incarnate Word Hospital	Fees	1,961,850	1933
St. Vincent's Hospital	Fees	1,441,017	1858
St. Mary's Infirmary	Fees	865,569	1877
Little Sisters of the Poor	Contrib.	740,762	1839
St. Anne's Home	Fees	545,735	1853
St. Louis University Hospital	Fees	419,186	—
Society of St. Vincent De Paul	Contrib.	412,916	1845
St. John's Hospital	Fees	400,000	1871
Mt. St. Rose	Other	380,377	1901
Catholic Family Service	Contrib.	378,996	—
St. Joseph's Hospital (St. Charles)	Fees	359,748	1891
Mother of Good Counsel	Fees	349,855	1929
St. Joseph's Hill Infirmary	Fees	334,843	1927
Our Lady of Perpetual Help	Fees	328,170	1957
Catholic Youth Council	Membership	258,480	1941
St. Agnes Home	Fees	234,622	1935
Cardinal Ritter Institute	Public	218,457	1961
St. Joseph's Institute for Deaf	Fees	216,408	1837
Convent of Immaculate Heart	Fees	215,924	1882
St. Louis U. School of Dentistry	Contrib.	213,892	—

Table 17 (continued)

Agency	Source of Funds	Income in 1965	Year Organized or Incorporated
Father Dismas Clark Foundation	Contrib.	141,543	1959
St. Joseph Home for Aged	Fees	140,344	1920
Frederic Ozanam Home	Public	131,643	1932
Child Center of Our Lady	Contrib.	130,032	1947
Father Dunnes' Home	Other	109,698	1906
Catholic Charities Administration	Contrib.	106,915	1912
Catholic Ch. Dept. of Children	Contrib.	102,132	1928
St. Joseph's Home for Boys	Contrib.	90,931	1835
German St. Vincent Orphans	Contrib.	88,168	1850
Convent of Good Shepherd	Public	76,101	1849
Guardian Angel Settlement	Contrib.	74,991	1911
Cath. Women's League, Day Care	Contrib.	73,481	1917
Father Jim's Home	Contrib.	64,533	—
Stella Maris Day Care	Contrib.	49,383	1845
St. Elizabeth's Day Nursery	Contrib.	48,492	1915
Mercita Hall	Contrib.	48,190	1835
Catholic Women's Association	Contrib.	46,770	1914
Catholic Charities Villa Maria	Fees	45,114	1956
St. Frances Home	Contrib.	40,625	1871
Msgr. Butler Neighborhood Center	Contrib.	36,758	1938
Sacred Heart Villa Nursery	Fees	35,370	1940
Catholic Charities of St. Charles	Contrib.	24,472	—
St. Patrick's Day Nursery	Contrib.	11,941	—
Helpers of the Holy Souls	Contrib.	6,856	1853
Total		$44,205,120	

Organizations for which income information was unavailable: Archdiocesan Committee on Human Rights; Bureau of Information; Camp Mater Dee for Girls; Catholic Guidance Center; Catholic Hospitals Association; Catholic Knights of America; Catholic Lawyers Guild; Catholic League; Catholic Physicians; Catholic Radio & Television; Catholic Women's College Club; Central Union of America; Cheer Club; Christian Family Movement; Council of Catholic Men; Council of Catholic Women; Daughters of Isabella; De Andreis Girl's Club; Don Bosco Camp for Boys; Father Dempsey's Charities; Hotel Alverne; Kelping Society; Ladies of Charity; Legion of Mary; Loretto Convent Pre-Kindergarten; McAuley Hall; Papal Volunteers for Latin America; Queen's Daughters' Home; St. Charles Home for the Aged; St. Louis Newman Foundation; Ursuline Child Garden; Webster Groves Reading Center; West Pine Reading Center.

CATHOLIC ORGANIZATION OF
SOCIAL WELFARE AGENCIES

Catholic social welfare is under the direction of the Archbishop of St. Louis. However, Catholic Charities and the United Fund are influential in many established programs to assist more than 400,000 Catholics in the city and county (*Yearbook of the Archdiocese of St. Louis, 1967:* 158). Of this number about 163,000 are in the city. Less than ten percent of the city's Negroes are Catholic, but Negro parishes have the fastest growth rates (Interview by author with Rev. J. A. McNicholas, September 8, 1967, St. Louis, Mo.).

Catholic Charities, a member agency of the United Fund, was established in 1912 and is a federation of forty-one agencies. Although having delegated planning and budgeting responsibility to the United Fund, Catholic Charities does consult with its member agencies on budgets, staff and services (Interview with Rev. McNicholas, 1967). More than three in four of these agencies are also members of the United Fund. There is less control over member agencies which are not members of the UF. No fund raising campaign is conducted for the Catholic Charities' agencies, but it does quietly solicit contributions. Previously, it was more autonomous from the UF. Allocation of UF funds to Catholic agencies is done only with informal consultation by Catholic Charities' staff.

As Table 17 shows, the estimated total income of Catholic social welfare institutions in St. Louis exceeded $44 million in 1965 (Special tabulation of data obtained from the HWC 1965 Expenditure Study).

Table 18

DATE OF ORGANIZATION OF SELECTED
CATHOLIC INSTITUTIONS, 1965

Year	Number of Agencies
Before 1861	11
1862–1899	10
1900–1914	6
1915–1929	9
1930–1944	11
1945+	10
Total	57

Community Service Directory (St. Louis: Health and Welfare Council of Metropolitan St. Louis, 1965).

This tabulation is based on fifty-two out of eighty-nine organizations for which information was available. Of the fifty-two agencies with incomes reported, thirty-two are members of the United Fund and their total income alone was nearly $34 million. Of this amount, $1.5 million is from the United Fund. The major source of income is from fees or membership dues, and the second largest is contributions. Funding from tax supported sources is increasing—particularly in the form of payments for children in

Table 19

BUSINESS AND RESIDENCE OF BOARD MEMBERS
OF CATHOLIC CHARITIES, 1966

Name	Business	Residence	Postal Zone
Bauman, G. Duncan	Globe Democrat	Clayton	05
Connors, Lawrence	Int'l. Assoc. of Machinists	County	23
Corley, Harry E.	Yates Wood & Co. Investments	St. Charles	
Ditch, Rev. R. G.	Catholic Charities	City of St. Louis	17
Fox, John	Mercantile Trust	U. City	30
George, J. Edwin, KSG	George Paint Co.	City of St. Louis	08
Gummersbach, Mrs. L. H.		U. City	30
Harrington, Harry F.	Boatmen's Bank	Ladue	24
Heneghan, Geo. E., KSG	Attorney	Clayton	05
Lamy, Charles S.	Investment Broker	Ladue	24
McMillian, Theodore	Judge	City of St. Louis	15
McNeela, Sister Mary Thomas, D.C.			
McNicholas, Rev. J. A.	Catholic Charities		
Medart, Mrs. J. R.		Ladue	24
Meier, George P.	John J. Meier Co.	Webster Groves	19
Miller, Rt. Rev. J. W.	Catholic Charities		
O'Sullivan, Mrs. Geo.		City of St. Louis	12
Richards, Roland W.	Laclede Steel	City of St. Louis	08
Ritter, Joseph Cardinal*	Archbishop of St. Louis		
Sassin, Mrs. Edward F.		City of St. Louis	09
Slattery, Rev. Robert P.	Catholic Charities		
Sullivan, Rev. L. A.			
Switzer, Mrs. Fred, Jr.		Clayton	05
Walsh, Mrs. Edward J., Jr.		U. City	30
Walsh, William D., KSG		U. City	30

* Deceased.
KSG = Knight of St. Gregory.
Annual Report/1966, Catholic Charities of St. Louis.

boarding homes, persons in nursing homes, and hospitals. In addition, federal grants to agencies have risen since 1964.

Table 18 shows the age of Catholic agencies. Almost one-fifth of the agencies were organized or incorporated before the Civil War, about 47 percent prior to World War I, and 18 percent established since the end of World War II. This is clearly a low rate in the formation of new agencies.

The Board of Catholic Charities supervises many agencies listed in Table 17. The names, business ties, and residence of the members of this board are given in Table 19.

Both businessmen and clerics are on the board; clerics in addition serve a triple role of priest, board member and staff. It is commonplace for clerics on boards also to be executives of agencies. For example, Messrs. Miller, Slattery and McNicholas, besides being officers of Catholic Charities, are also its top administrators. Only Fox and Harrington are

Table 20

EXECUTIVE OFFICERS OF UNITED FUND AGENCIES UNDER CATHOLIC SPONSORSHIP BY TOTAL BUDGET AND BY CLERICAL AND LAY AFFILIATION

Agency	1965 Budget (in Thousands)	Clerics	Laity	Total
Catholic Charities	$ 677	4	3	7
Society of St. Vincent De Paul	413	—	6	6
Catholic Youth Council	258	3	1	4
Cardinal Ritter Institute	218	1	3	4
St. Joseph's Institute for Deaf	216	4	—	4
Frederic Ozanam Home	132	3	2	5
Child Center of Our Lady	130	3	2	5
St. Joseph's Home for Boys	91	4	—	4
German St. Vincent Home	88	—	7	7
Convent of the Good Shepherd	76	—	4	4
Guardian Angel	75	4	1	5
Catholic Women's League	73	—	7	7
Stella Maris Day Care	49	4	—	4
St. Elizabeth Day Care	48	1	5	6
Mercita Hall	48	4	—	4
Msgr. Butler Neighborhood Center	37	—	5	5
Totals	$2,629	35	46	81

Excluding hospitals and two agencies for which information was not available. Data from the Expenditure Study and agency reviews completed by the Health and Welfare Council.

from Civic Progress companies. In spite of having well-to-do people on its board, locally based, large-sized corporations are underrepresented. Location of board members in the most economic advantaged areas of the city presents a different residential distribution than for affluent Jews or Protestants who are more concentrated in the county.

Further evidence of clerics being in influential positions is seen in Table 20. Priests and sisters had only 43 percent of the officerships, but agencies which had a majority of clerical officers were in control of almost 65 percent of the money to be spent.

In some agencies there are two boards: governing and lay. St. Joseph's Hospital in Kirkwood has this type of agency government and is shown below (Evaluation of a Capital Fund Project Proposed by St. Joseph Hospital. Prepared by Metropolitan St. Louis Hospital Planning Commission, April, 1966.):

Governing Board (Sisters of St. Joseph)

Sister Anita Louise	President
Sister Margaret Eileen	Vice-President
Sister Carol Joseph	Secretary-Treasurer
Sister Domitilla	Member-at-Large
Sister Aguina	Member-at-Large

Lay Board

W. Donald Dubail	President
David L. Jones	Vice-President
Ellis Phillips	Secretary
John Warner	Member
Donald Gunn	Member
Henry Buhr	Member

Not only were Sister Anita Louise, Sister Margaret Eileen and Sister Carol Joseph officers of the board, but the head cadre—respectively: Administrator, Assistant Administrator and Controller.

Even though the clergy are visible perforce of habits in both staff and lay roles, many non-Catholics are served by Catholic agencies. The Convent of the Good Shepherd for delinquent girls, operated by the Sisters of the Good Shepherd, is nonsectarian in its intake because care of delinquent children is defined as a public function. However, twenty-four hour care of children in Catholic homes is restricted to Catholics, and is rationalized on the basis they could be accused of proselytizing if they accepted Protestant children.

Not all Catholic social welfare in St. Louis centers in Catholic Charities. Coterminous with Catholic Charities, founded by Msgr. John J.

Butler, is Father Dempsey's Charities. The latter developed nonprofession-
ally and the former professionally according to social work standards.
Antithetical systems evolved from their efforts, and both systems were
blessed by the Archbishop.

The Archbishop collects money for charitable causes. The Expansion
fund, some of which goes to charities, is obtained through free will offer-
ings. In addition, the Cardinal initiated Tithing-for-the-Poor to aid less
advantaged parishes ("Cardinal Ritter Announces Tithing-for-the-Poor
Plan," *Globe-Democrat,* January 21, 22, 1967). Some of this money has
been used to match money available from the Office of Economic Opportunity (Interview with Rev. McNicholas, 1967). Tithing-for-the-Poor
under the direction now of Bishop Gottwald is administered by a board
of six priests who have no training in social work.

Announcement of the tithing plan came about ten days after publicity about many Catholic churches being gaming places for raffles and
bingo games. This source of income for charitable, as well as for other purposes, was threatened by Col. Edward L. Dowd, president of the St.
Louis Board of Police Commissioners. He said gambling is gambling
even in a church and could lead to hoodlum activities. Cardinal Ritter also
noted that gambling violates church rules (John R. Bell II, "Charity
Group Members Split on Gambling," *St. Louis Post-Dispatch,* January 12,
1967). However, the practice remains widespread in the St. Louis area
according to Bell, the *Post-Dispatch* reporter.

Innovations occur outside the established social work institutions.
The Rev. John Shockley, pastor of St. Bridget's church, is pioneering in
the field of housing for low income people. Because of his efforts and those
of Father Joseph M. Kohler, his former assistant, the Bicentennial Civic
Improvement Corporation (a private, nonprofit corporation) was established to obtain capital to buy tenement houses in the Mullanphy Area
of the Near North Side (Richard M. Jones, "Rebuilding a Neighborhood,"
St. Louis Post-Dispatch, October 9, 1966). These houses are purchased,
remodeled and sold at low interest rates by the Corporation. Because more
money was needed to expand the program, approval was sought from
the Health and Welfare Council to have a limited campaign to raise $100,-
000. Union Electric had made a challenge grant of $33,000. The Council
subsequently approved, but raised objections to unannounced visits by
the clergy to see if the new homeowners were behaving themselves and
to the requirement of husband and wife publicly expressing affection. The
term "worthy of assistance" is considered patronizing by the Council
staff—placing this program in the category of charitable giving, and
hence likely to defeat the objective of rehabilitation and fostering self-
respect (Memorandum to Bicentennial Civic Improvement Corporation
Review Committee, Health and Welfare Council of Metropolitan St.

Louis, August 5, 1965). This is indicative of a professionalizing influence, and legitimation of a secular nature.

Another example of quest for a mission was the decision of the St. Louis Catholic Cathedral's parish study committee to survey the area around the Cathedral to determine what needs might be served. Daniel L. Schafly, Jr. was on the committee and his father contributed $1,000 to pay the Health and Welfare Council for research services. Mr. Schafly, Sr. is a blue ribbon member of the Board of Education, Chairman of the newly appointed lay board to St. Louis University, and on the board of Catholic Charities.

The Council's study is of interest because it assessed social welfare needs for Catholics in an area which includes others besides Catholics ("The St. Louis Cathedral Parish Area, A Study of the People, Physical Conditions, Problems and Needs," Health and Welfare Council, November 30, 1966). No attempt was made to involve non-Catholics except as sources of information. This study shows the elitist nature of welfare organization starting with a small group funded by a civic notable. Moreover, the Council engaged in Catholic social welfare planning which is a role Catholic Charities sought for itself in 1965.

Since Catholic parishes wanted to become involved in the War on Poverty, Catholic Charities proposed that it be given responsibility to develop antipoverty programs. This recommendation was turned aside by the United Fund and Health and Welfare Council because this type of planning by Catholic Charities could lead to further fragmentation of planning. A new director of the Health and Welfare Council was starting his tenure at the time. The incoming board president was to be G. Duncan Bauman; the past board president was John Poelker (Comptroller of the City of St. Louis). The Treasurer of the Council was David Q. Wells, all of whom are Catholic, except Mr. Wells whose wife is Catholic.

PROTESTANT ORGANIZATION
OF SOCIAL WELFARE AGENCIES

Protestants have no counterpart to the Jewish Federation and Catholic Charities in St. Louis. The Metropolitan Church Federation lacks the funding and validation to serve this function. The Community Fund established in 1922 included Protestant and nonsectarian agencies, but the Jewish and Catholic federations rejected union with the Fund at that time (Levy, 1928:62–63). Hence, up until the 1930s the fund represented Protestant interests exclusive of Jewish and Catholic. With the "merger" then it continues to be a champion for Protestant agencies.

Table 21 indicates that about $28 million was obtained for operating

Table 21

PROTESTANT AGENCIES BY MAJOR SOURCE OF FUNDS,
TOTAL INCOME AND YEAR ORGANIZED

Agency	Major Source of Funds	Income in 1965	Year Organized or Incorporated
St. Luke's Episcopal-Presbyterian Hospital	Fees	$ 5,619,537	1866
Lutheran Hospital	Fees	4,976,077	1858
Deaconness Hospital	Fees	3,890,808	1889
Missouri Baptist Hospital	Fees	3,406,937	1884
YMCA	Fees	2,195,425	1853
Christian Hospital	Fees	1,918,504	1910
Salvation Army	Contrib.	1,022,143	1880
YWCA	Fees	465,727	1905
Gatesworth Manor	Fees	400,000	1961
Home of the Friendless	Contrib.	312,000	1853
Good Samaritan Home for Aged	Contrib.	310,159	1856
Railroad YMCA	Fees	262,761	—
Christian Old Peoples' Home	Fees	241,201	1914
St. Louis Christian Home	Fees	239,524	1889
Salvation Army Booth Memorial Hospital	Contrib.	199,593	1898
Methodist Children's Home	Contrib.	192,543	1864
Missouri Baptist Children's Home	Contrib.	176,306	1886
Emaus Home	Fees	150,000	1893
Grace Hill House	Public	147,527	1906
Lutheran Convalescent Home	Fees	140,388	1920
Evangelical Children's Home	Contrib.	136,797	1858
Lutheran Altenheim	Fees	115,949	1906
General Protestant Children's Home	Other	110,891	1877

health and welfare services of forty-three agencies under Protestant auspices in 1965. The eighteen UF agencies included received more than $24 million of which $1.5 million came from the United Fund allocations— the same amount received by Catholic Charities. Most of the $28 million is from fees.

Table 21 (continued)

Agency	Major Source of Funds	Income in 1965	Year Organized or Incorporated
Lutheran Children's Service	Contrib.	109,124	1863
Epworth School for Girls	Contrib.	109,034	1909
Presbytery of St. Louis	Contrib.	106,993	1959
Lutheran Children's Home	Contrib.	101,399	1863
Kingdom House	Contrib.	91,770	1902
Caroline Mission	Contrib.	64,390	1941
Lutheran Boarding Home	Fees	63,906	1906
Fellowship Center	Public	62,738	1943
Lutheran Mission	Contrib.	61,809	1899
Camp Auroa	Contrib.	58,000	—
Plymouth House	Contrib.	39,973	1960
Greeley-Presbyterian Community Center	Contrib.	34,100	—
St. Stephen's Church	Contrib.	31,636	1886
Salvation Army St. Charles	Contrib.	20,177	1939
Christian Counseling & Guidance	Fees	18,900	1961
Metropolitan Service Association	Contrib.	17,808	—
Pastoral Counseling Service	Fees	6,880	—
Metropolitan Church Federation	Membership	5,404	1909
Diocese of Missouri Protestant Episcopal	Membership	4,391	—
Missouri Episcopal Federation	Membership	3,149	—
Total		$27,642,378	

Organizations for which income was unavailable: Centenary Counseling Service; Christ Church Cathedral; Christian Women's Benevolent Association; Council of Lutheran Churches; Episcopal City Mission; Lutheran Community Center; Lutheran Human Relations Association; Lutheran Special School; Memorial Home; Neighborhood Health Center; Sunshine Mission; Tower Grove Manor; United Church Women of St. Louis, and Youth Counseling Service. The source of the expenditure data is the Health and Welfare Councils' study of income and expenditures.

Table 22 shows that 70 percent of these agencies were established prior to World War I, and less than 20 percent have been formed in the last twenty years. The age of the agencies is comparable to that of the Catholics.

The Metropolitain Church Federation, established in 1909, periodi-

Table 22
DATE OF ORGANIZATION OF SELECTED
PROTESTANT AGENCIES, 1965

Year	Number
Before 1861	5
1862–1899	17
1900–1914	11
1915–1929	1
1930–1944	4
1945+	9
Total	47

Community Service Directory (St. Louis: Health and Welfare Council of Metropolitan St. Louis, 1965).

cally attempts to become a member of the United Fund (Interview by author with Dr. O. W. Wagner, September 18, 1967, St. Louis, Mo.). The need to coordinate and plan for expansion of Protestant social services is the justification. Also, MCF contends that its constituency is concerned that Catholics receive more than their share of the community's contributed dollars. The Health and Welfare Council opposed membership of MCF in the United Fund because it would mean another welfare planning body. This position assumes that the Council already assists church social welfare programs. Dr. Wagner, the head of MCF, however, argues that the Council looks upon Protestant social work as substandard, which he says is a naive position. He thinks that if MCF could circumvent HWC, which he refers to as the right arm of the UF, the latter would grant membership primarily because there is mounting criticism of undue Catholic influence in the United Fund.

According to MCF's director, the Protestant power structure is composed of Civic Progress influentials. Moreover, Dr. Wagner sees a graduation from the board of the Downtown Young Men's Christian Association to the Metropolitan board of the YMCA, and anyone who is somebody moves from there to the Boy Scout board which is the prestige pinnacle of the agency boards. If the Protestants are so powerful, why is MCF not a member of the United Fund? The MCF director explained that he has not played the game of power politics and that he does not intend to now. Another committee has been appointed by MCF to present its case to the UF, but Dr. Wagner says that he does not want to use one of the Civic Progress notables who said to let him know when MCF's application is resubmitted.

The administrative committee of MCF is its central decision making body, but there are no business elites from Civic Progress corporations represented. Many church people, Dr. Wagner said, would resent placing someone on this committee simply because of being from the power structure.

MCF publicly opposed the formation of the UF because it would be undemocratic, representing only big business interests, and its board would be self-perpetuating.

There are various intradenominational federations of Protestant churches: Presbytery of St. Louis, Episcopal Diocese of Missouri, Council of Lutheran Churches, United Church of Christ, and others. These different church bodies have their own departments for social welfare planning. The head of MCF maintains that these different church bodies would welcome the central coordination which MCF could provide if subsidized through the UF. Presently, these Protestant federations deal directly with the UF in negotiations concerning agencies under their auspices which are also members of the United Fund.

Concerning social welfare ideology, MCF's research director, Dr. Raymond Schondelmeyer, spoke of "crisis centered activists" and the "heaven glancing traditionalists." The former want the church in the vanguard to save cities from the ravages of poverty and crime, and the latter believe the church should not involve itself in politics and welfare. This is reflected in the differences between the churches which stayed in the slums and those which took flight to the suburbs. Schondelmeyer (1967) advocates a vigorous attack on the conditions that cause poverty and slums.

> The Church is concerned about man and the expression of his life as an individual and as a member of his society. The city exists to serve him and afford him opportunity for freedom—social, political, and economic. The action of the church must affirm the dignity and value of the city's people. Institutions and instruments of the metropolis must be the people's servant and not their master.

Schondelmeyer is speaking as a "revolutionary" and perhaps in a sense the clerical role which embraces this ideology does stand for something which is unique and apart. He questions the status quo, and implies that the ruled should be the rulers. Schondelmeyer further states (1967):

> Protests and stands on social issues, even though they are controversial, are a very essential role of the Church. It is imperative that a number of issues be kept before the public and discussed openly so that these issues may be clarified and that the public welfare best be served. It is for the same reason the church supports the right to dissent.

Again he is stating a position which would make the clerical role distinctive.

The Protestant Ethic controls the thinking of many people, says Dr. Schondelmeyer (Interview by author with Dr. R. Schondelmeyer, April

3, 1967, St. Louis, Mo.). They see salvation through work—one's calling. Because often their incomes are low, they favor minimal assistance for those on relief. In order to encourage individuals to support themselves, relief grants must be unattractive.

The YMCA is used as an index of Protestant influentialism because of the Y's high social rank in the social agency system. Comparing employment of the board members of the YMCA, Jewish Federation, and Catholic Charities, the relative power of these sectarian bodies may be inferred (see Table 23). The YMCA leads by having a total of fourteen

Table 23

BOARD MEMBERSHIPS IN SECTARIAN AGENCIES BY EMPLOYMENT
IN CIVIC PROGRESS COMPANIES, 1966

| | Board Memberships | | | |
Company	Catholic Charities	JF	YMCA	Total
Anheuser-Busch	—	—	—	—
Bank of St. Louis	—	—	—	—
Boatmen's Bank	CEO	—	Pres.	2
Brown Shoe	—	—	Pres.	1
Clarence Turley Inc.	—	—	—	—
Falstaff	—	—	—	—
First National	—	VP	CEO; Pres.	3
General American Life	—	—	CEO	1
General Steel	—	—	—	—
Granite City Steel	—	—	—	—
Interco	—	—	CEO	1
Laclede Gas	—	S'LMN.	—	1
May Department Stores	—	—	—	—
Mercantile Trust	Pres.	—	—	1
Monsanto	—	—	VP	1
Pet Co.	—	—	Retd. VP	1
Ralston Purina	—	—	VP	
			Retd. Treas.	2
Price-Waterhouse	—	—	CEO	1
Reinholdt & Gardner	—	—	—	—
St. Louis-San Fran. R.R.	—	—	—	—
St. Louis Union Trust	—	—	VP	1
Southwestern Bell	—	—	CEO	
			Retd. VP	2
Totals	2	2	14	18

Board lists of agencies in addition to Polk's St. Louis City and St. Louis County Directories.

of its forty-one board members working for or retired from Civic Progress companies. Five members of Civic Progress are on the board of the YMCA: Chambers, Goodson, Hickok, Jackson, and Peirce. Thus, Civic Progress corporations are represented eighteen times and the YMCA has nearly 78 percent of these. Catholic Charities, with twenty-six board members, has only Harrington from Civic Progress on its board of directors, although it does have a bank president, but he is not in Civic Progress. The Jewish Federation, with fifty-two members, has neither Edison nor May on its board, although these are chief executive officers of their companies. Thus it appears from this tabulation that Protestants —through the United Fund—have considerably more influence in welfare decisions than might be appreciated only from a superficial examination of the materials.

For the funding of special programs, the YMCA, Salvation Army, Grace Hill House, The United Church of Christ Neighborhood Centers, and Kingdom House—all of which are sectarian or church affiliated agencies—have taken funds from the Office of Economic Opportunity through the Human Development Corporation. This has been carried out under the close, and at times critical, supervision of the United Fund. United Fund staff point to the possible neglect of ongoing programs; business influentials are wary because of the uncertain funding from OEO. If withdrawn, the UF may be called upon to raise money to continue these programs. However, because of community needs, the United Fund has approved acceptance of public funds. Of the forty-three agencies listed in Table 20, thirty have received public subsidies. Organizations lacking subvention have smaller budgets. A tabulation of OEO money received by UF agencies in 1966 shows a total of $1.7 million. Of the thirteen agencies receiving these funds, eight are under sectarian auspices.

Dr. James Spivey (Interviewed by the author on March 24, 1967 in St. Louis, Mo.), the executive director of the St. Louis Presbytery in an interview, said concerning the church's responsibility in welfare:

> The needs have become so marked that it is too much for the churches to handle, and has to be taken over in part by government. The basic motivation came from religious heritage, but it is generally beyond their resources now. Much of it had to be relinquished to get the job done. Very few of the denominations administer just to themselves. The Mormons are about the only group who have extensive services for their members. Catholics and Protestants do not limit participation but give services with no strings attached. Much has been given up to the secular institutions, and the question is do we try to take it back? Do we try to influence the agencies which are doing social work? Or do we do some combination? Perhaps the real role is to pick up on things that the public has not done.

Dr. Spivey is probably speaking for many Protestant leaders—particularly reflecting the viewpoints of economic influentials who support not only the Presbytery's health and welfare programs but who determine social policy in the community. His comments concerning control of church welfare are similar to those who run the United Fund. He says, "Who pays for it, controls it." This is the reason why the Presbyterians have tried to avoid having their institutions directly subsidized. He referred to the Presbytery's action forbidding the acceptance of federal funds for its own community service programs. Church property, however, can be used. Dr. Spivey said this action by the Presbytery might be reversed. They have gotten around this by having money go to "paraparochial" agencies which have been set up sometimes jointly by several denominations. The staff are ministers, but they do not have an administrative linkage with their parent church body. The Mid-City Community Congress and Exit are examples of ambiguous agencies which are neither numerous nor influential. He sees the paraparochial institution as a way to carry out innovative programs "without being bought off or shut up," because under church auspices, they would be under more pressure. This leaves the church free to criticize government whereas with direct funding it would not be able to do so. The setting up of Gatesworth Manor, a rather expensive home for the aged, as a separate legal entity, from either the Presbyterian or Episcopalian churches (the joint founders), is another example of paraparochialism. The tax-exempt status of this institution is being questioned by the city on the grounds that the costs for care are as high or higher than some of the commercial homes. Dr. Spivey does not think they will lose their tax-exempt status. He thinks instead that they should make a gift to the city since such an institution realizes benefits from police, fire protection, and other city services. He is opposed to taxes because he says, "The city could tax the private agencies out of business." Dr. Spivey is interested in serving the rich as well as the poor, but favors more equity. He is not concerned about government intervention; our system of government, he contends, is strong enough and with sufficient diversity to preserve freedoms. He would like to see more central direction in the metropolitan area. He is concerned about the fragmentation of the churches, and ministers getting into social welfare programs for which they are not professionally trained.

OVERALL ORGANIZATION OF
SECTARIAN SOCIAL WELFARE

In the St. Louis area, nearly $82 million a year is spent on sectarian health and welfare services, about two times more than for private nonsectarian.[8] Of this amount, more than $60 million is from fees as compared to only $8.3 million from contributions. The United Fund's allocations, included in the above figures on contributions is about $3.8 million. Hospital expenditures overshadow the others, representing 79 percent of the eighty-two million dollars.

Although to the casual observer financing and organization of sectarian services appears pluralistic and somewhat anarchical, the facts presented in this chapter indicate otherwise. Amid the seeming organization pluralism, decisions which influence the policies and programs of hospitals, family agencies, children's institutions, and other such organizations are imperatively coordinated and dominated.

The organization of sectarian welfare is tied to the same upper stratum of business influentials who control nonsectarian health and welfare service systems of St. Louis. At this level, social class substitutes for religion and ethnicity (Baltzell, 1966:63).

Business influentials active in Jewish, Catholic and Protestant agencies are also the same ones who make high level social policy decisions. Although Protestant leaders are more frequently from the largest locally based industrial corporations, Jews and Catholics are in this upper stratum. The latter have not penetrated the industrial empire of St. Louis as much as Protestants, but Jews and Catholics are represented in powerful commercial and banking establishments. This fits Baltzell's theory (1966:71) insofar as Catholics and Jews have high economic power even though their social status is lower. St. Louis Catholics tend to be of higher status because many of their antecedents were of French and German origins who never had as low a status as Catholics from other countries (Consultation by author with David J. Pittman, November 20, 1967, St. Louis, Mo.). It is questionable that the representative establishment of which Baltzell speaks, has been reached since Jews and Catholics are under represented in large business and the latter are dominant in political power in the city.

Civic Progress, Inc., constituted with these above-mentioned influentials, through the United Fund and key sectarian organizations such as the Jewish Federation and Catholic Charities, makes strategic decisions about sectarian health and welfare services. As previously indicated, Civic Progress, Inc. created the United Fund and dominates it. Limits are set on expansion of sectarian services through both private and public funding. Civic Progress members determine the UF goal which accordingly controls UF agency budgets. A monopoly of corporate solicitations regulates the amounts raised by campaigns outside the UF. The hospital field, which represents almost eight in ten dollars expended for sectarian services, has long been an interest of civic influentials. Although currently the UF allocates less than one million dollars annually for sectarian hospital care, the UF through the Hospital Planning Commission regulates expansion of this field. It has indirectly financed the Hospital Association concerned with promoting efficient hospital management. Other sources of funding such as increased fees and insurance rates, depreciation allowed as operating expenses, Medicare and federal grants have relieved Civic Progress influentials of fiscal responsibilities in the hospital field. However, no evidence is reported that the influentials fought the additional or new sources of funding. In fact, General American Life serves as a fiscal intermediary for Medicare. The UF is willingly phasing out of financing hospital care because the medically indigent load can be handled outside private philanthropy. Moreover, the United Fund along with Blue Cross approved depreciation allowances. This of course caused hospital care costs to rise.

Support of sectarian welfare programs comes through "taxes" of the corporations, fees paid for services, and of course, private and public insurance programs. The UF urges that agencies increase their income from fees which can result in corresponding reductions in UF allocations. This is consistent with the principle of getting the best mileage possible from the giver's dollar. Because money from the UF is often "last" money, it is essential for many agencies in order to continue their operations. Welfare programs are not restricted to the worthy poor, but are for everyone in the community who is able to take advantage of them. Both the affluent and less affluent are paying the welfare costs. Payroll deductions for charities, set by formulae dictated by the United Fund and geared to increase with wages, or through governmental taxes, limits growth of social service systems. Corporate gifts are from corporate profits; executive and employee giving from salaries and wages. Corporations, executives and employees have their interests to protect. The corporations need profits for new expansion and for dividends, and to pay executive and employee salaries. Executives have their own health, pension and stock purchase plans. Likewise, employees have their pension and other health and welfare programs for which there are deductions. Hence, keeping expansion rates low for sectarian or nonsectarian welfare, either from churches or from the UF, is virtually an automatic process. Likewise, public social services are depressed to keep taxes down.

Two ideological currents merge: the Protestant Ethic and the Corporate Ethic. The first emphasizes frugality, hard work, and personal responsibility. In addition, it includes the principle of worthiness which is related to personal responsibility. Those addicted to narcotics, criminals, and people who do not try to help themselves are largely outside the ken of community service. The second current is manifested in welfare which flows from the corporations. A basic assumption is that the burden should be distributed over the masses (e.g. "Everybody gives and everybody benefits"). This is analogous to individual responsibility. Besides the welfare programs which can be generated through the cooperation of the large corporations, the Corporate Ethic sanctions the use of governmental funds along sound business lines. The union of the Protestant and Corporate Ethics is functional because it protects corporate profits and executive salaries from being threatened by excessive taxes whether they be privately or publicly imposed upon the upper classes. Moreover, these ethics foster the wholesome image of the civic directed corporations and their civic influentials.

There is uncertainty as to how particular decisions will be made affecting sectarian welfare programs. Some matters are decided at lower levels in unpredictable ways. However, this is consistent with the model of the corporation which does delegate authority. Nevertheless, decisions are reviewed by men at the top; they have the power to intervene, to select what they think is important enough not to leave to the vagaries of the system.

Excluded from the Protestant Establishment, as represented by the United Fund, are other social service systems. Some of these are avantgarde, protest agencies which are not respectable enough to be admitted to the United Fund; others are respectable but do not think they can obtain as much money for their causes if they align with the United Fund. The multiassociational character of sectarian social service standing apart and opposed to the United Fund is probably useful in that charity based on religious concern is dissipated. Elites do not raise as much money as they might otherwise have to do. There is less criticism of civic notables when the system is somewhat fluid. Moreover, as long as solicitations of large corporations are restricted to the United Fund, they are not dismayed by other drives which for the most part are headed by lower stratum people.

CONCLUSIONS

Sectarian social welfare is controlled by the Protestant Establishment which has incorporated Jews and Catholics into the total secular power structure represented in the United Fund which makes basic social policy decisions regarding the organization of social welfare in St. Louis. What is not part of the Establishment may be functional for the maintenance of the Establishment and is not likely to modify the status quo because its support either comes from the corporation or the government.

Dominance by Protestants is due to their control of corporate wealth. Jews and Catholics, although penetrating large business, have not achieved the economic dominance of Protestants.

Religion and sectarianism are submerged in social class at the top. The organization of health and welfare programs derives from an interrelated system of reciprocities based on economic and philanthropic "back scratching." Although Jews, Catholics and Protestants have varying degrees of autonomy over welfare programs, when it comes to the United Fund sectarian interests are depressed even though underneath they remain viable.

Secularization is seen in the UF's principle of services being open to all. Utilization is a different matter with many agencies having de facto segregation of religious groups. Acceptance of public funds tends to break down the sectarian boundaries.

Control exerted over sectarian service systems is manifested by the ages of agencies and phasing out in the hospital field. Reluctance to support new agencies through private philanthropy and holding down expansion of existing agencies except through spreading the costs using fees or tax funds are indicative of putting profits and executive salaries first.

The Protestant Ethic and Corporate Ethic both provide the ideology which is widely accepted by the masses. The Protestant Ethic keeps welfare expenses minimal as applied to the poor. The Corporate Ethic is more generous—welfare is for the affluent and nonaffluent alike. Hence, programs which the nonpoor can benefit from are more favored. This can be seen in the favored position of the Boy Scouts, YMCA, and hospitals.

Both Protestant and Catholic clergy speak for social justice. Education and employment programs like those sponsored by the Jewish Employment and Vocational Service are promises of the new welfare but still largely for the worthy group. However, it is perhaps too soon to say if these efforts will be more than charity in a new form. If the people educated and trained are employed at wages which do not yield sufficient family income for adequate levels of living, then social justice is compromised.

Part Four

POLITICAL

ORGANIZATION OF

SOCIAL POLICY

Chapter 8

Mayor's Office:

Brick and Mortar

The emphasis, thus far, has been on aspects of private power. Instances of lines of power and action vis-à-vis Civic Progress and public agencies have been noted. However, up to this point dominance of private over public power has not been fully delineated. Now, the time has come to beard the pluralist lion. According to the pluralist view, compartments of business, social welfare, government, and education are relatively independent.

In order to demolish the pluralist position one must be able to demonstrate that major social policies are governed by established private and public structures paramountly interpenetrated by private corporate power.

This chapter shows that the Mayor's Office is basically interpenetrated by an interorganization of private agencies whose lines of power stem from large locally based corporations. Besides structural aspects, specific transactions between the Mayor's Office and influential private agencies are cited.

City Hall in St. Louis refers to that network of functions and functionaries legitimated by the City Charter (August 29, 1914). The three major divisions of City Hall are: the Mayor's Office, the County Offices and the Board of Aldermen. Neither the Mayor's Office nor the County Offices has control over the other. The Board of Education, structurally independent of city government, is supported in part by a district tax rate determined by referendum. The Police Commission, with the exception of the mayor, who is a member, is appointed by the governor.

Civic Progress manages key public institutions in terms of social policy. Establishment agencies have their gyroscopes instrumented by private ideology, being basically oriented to creating conditions conducive to the pursuit of invidious pecuniary gains.

The Establishment, consisting of both public and private agencies, may be considered as a Civic Progress conglomerate with three portfolios of holdings: Civic Progress preferred private; Civic Progress nonpreferred private, and Civic Progress nonpreferred public. Outside these portfolios are nonestablishment agencies which may be sucked into this conglomerate.

Opposed to the Establishment are protest groups seeking to work within the limits of sanctioned private and public power, and still others convinced that violence is an appropriate response to exploitation based on gross inequities in the allocation of scarce resources (Lipsky, 1967. Also see Horowitz and Liebowitz, 1968:281–296). For the most part these are black and student organizations imperfectly socialized to private ideology. Nonestablishment organizations outside the purview of this study are not particularly significant, and can be snuffed out in various ways, if need be, by legitimate State violence. The Mafia should not be classified as non-Establishment because of being interlaced with the Establishment itself; and depending upon a private ideology for the estimated thirty billion dollars which it takes from the economy annually.

Figure 5 shows the preferred private and nonpreferred public Establishment agencies.

Civic Progress, Inc. is central to the Establishment power configuration.[9] As already indicated, the economic wealth of St. Louis is well represented by Civic Progress corporations which contribute handsomely to the tax base; directorates of the corporations and major social agencies are interlocked, and their power is further organized through Civic Progress itself which serves as the executive committee for corporate influentials. Excluded from membership in the august "senate" are the St. Louis Mayor and the Supervisor of St. Louis County. However, they are invited to its meetings.

Dominance is manifested as follows: (a) Civic Progress controls through funding of voluntary and tax-supported agencies; (b) Interlocking directorates in their own corporations and in key agencies foster consensus consistent with the Civic Progress ideology, and the close network of interpersonal relations facilitate mutually advantageous reciprocities; (c) Civic Progress members have substantial investments in the community which cause their participation in the political organization of welfare through City Hall, and conversely their political activity augments those investments, and (d) Party loyalties as Democrats or Republicans are subordinated in the interests of corporate profits and civic progress, defined as interdependent by this Establishment.

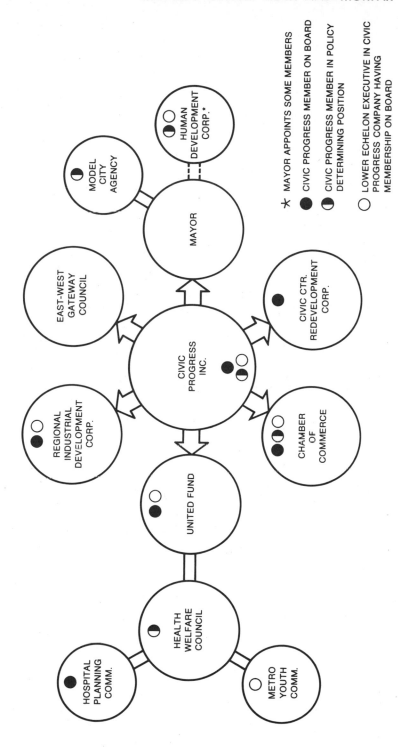

★ MAYOR APPOINTS SOME MEMBERS

● CIVIC PROGRESS MEMBER ON BOARD

◐ CIVIC PROGRESS MEMBER IN POLICY DETERMINING POSITION

○ LOWER ECHELON EXECUTIVE IN CIVIC PROGRESS COMPANY HAVING MEMBERSHIP ON BOARD

Fig. 5. Model of the Political Organization of Social Welfare in the City of St. Louis

MAYOR'S OFFICE

The Mayor's Office, the nerve center of City Hall, is an important part of the Establishment because it links various city departments, the state, and federal bureaucracy. Both mayor and governor are brokers for Establishment holdings. A network of reciprocities tie private and public sectors together, but taxes for the city and the remainder of the state come to a great extent from mammoth corporations and their employees. Although the Mayor's Office has been enhanced by new federal programs, it is transparent that the federal programs' success, just as with the United Fund, depend upon the cooperation of business influentials.

The agencies in Figure 5 are coordinated through Civic Progress. As gatekeepers of the political process which formulates urban social policy, these elements of the total agency structure stand out as providing central channels for regulation of change. This private and public power structure is committed to maintaining a capitalistic system through the Protestant and Corporate Ethics with emphasis on private and public welfarism. Education, housing, health, recreation, and social programs are substituted for adequate income (Rainwater, 1967:40–43). Really comfortable income would mean lower profits; likewise, minimum or below minimum wages perpetuate exploitation. Guaranteed annual income or other income strategies may only reinforce existing inequities if per capita incomes remain paltry.

The Mayor's Office consists of the Board of Estimate and Apportionment plus the Board of Public Service (see Figure 6). The Board of Estimate and Apportionment is responsible for fiscal policies regarding the city. The mayor is in charge of working up the budget; the board of aldermen may cut but not increase appropriations recommended by the mayor. The members of the Board of Estimate and Apportionment are: the Mayor, Alfonso J. Cervantes; the Comptroller, John Poelker, and the President of the Board of Aldermen, Donald Gunn, all of whom are elected at large. These three political leaders are Democrats and Catholics. About one-third of the city's people are Catholic.

There is a board of aldermen composed of twenty-eight members who are elected from twenty-eight wards in the city. Only four of these aldermen are Republicans. The board of aldermen usually supports the mayor's programs.

The board of public service consists of heads of major departments of the city: health and hospitals, welfare, streets, public safety and others. Being able to appoint heads of departments gives the mayor some power even if regulated by civil service.

Voters

Board of Estimate and Apportionment

Mayor   Comptroller   President,
                                Board of
                                Aldermen

Board of Public Service

Director	Director	Director	President	Director	Director	Director
Parks	Health	Welfare	Board of	Streets	Public	Public
Recreation	&		Public		Safety	Utilities
&	Hospitals		Service			
Forestry						

Fig. 6. Mayor's Office, City of St. Louis.

Because the City of St. Louis is a county as well as a city, there are
county offices. The Mayor's Office and County Office group are described
by Banfield (1965:124) as making city policy and managing the patronage
respectively. The Mayor's Office, he says,

> . . . stands for "good government" and "civic progress." Its chief
> supporters are the newspapers, the downtown business interests, and
> the middle class wards. Its policies, however, are not always conserva-
> tive; in race relations, and even in fiscal matters it has often been
> innovative and progressive. The other set, the County Office Group,
> consists of county officials and some aldermen. They have little to do
> with policy. They are "politicians" in the narrow sense; their main
> interest is in getting and giving jobs and favors. Their base of support
> consists of Negroes, small businessmen, the politically active elements
> of organized labor, and the low-income wards.

The sheriff, collector of revenue, license collector, and recorder of deeds
constitute these county elective offices.

Banfield (1965:124–125) points out that the party organization is
controlled by elected committeemen who decide on the persons to run for
alderman in their wards. He states that:

> The County Office Group has enough patronage to tie the committee-
> men into a machine that might control the city. Instead of pooling
> their patronage to do this, however, the various elements within the
> County Office Group bid against each other for the committeemen's
> support. This makes it possible for the committeemen to maintain
> their independence (one who refuses to knuckle under to one element
> of the County Office Group cannot be well disciplined, for a com-
> peting element is always at hand with an attractive alternative) and
> in effect it guarantees that a diverse set of interests will be represented
> on the Board of Aldermen. . . . Consequently, the mayor is seldom
> confronted by a large stable opposition block within the board. . . .

One of the illuminating events in 1965 was the Democratic mayoralty
primary of Raymond R. Tucker, the incumbent, and Alphonso J. Cervantes, his redoubtable and quixotic opponent. It was a campaign filled with
verbal onslaughts and innuendos. Tucker stood for the forces of civic
progress. He had been mayor since 1953. The money and manpower of
Civic Progress were on his side. After four terms in office, the rebirth of
the downtown was assured, and the dollar harvest was at hand. The 630-
foot arch would be the glistening monument to civic progress. Cervantes was
linked by the newspapers to the less than respectable elements of the community: the steam fitters union, taxi cab interests and to various business
promotions which had made him a millionaire. From the ward to the presidency of the board of aldermen and finally to running against Tucker constituted a long period in which his own business interests and his political

aspirations were interlocked. Significantly, Cervantes had the backing of the "new Negroes" who had flatly rejected the Uncle Tom leadership of the Tucker camp (Streck, 1965:13).

Tucker lost. There were many reasons, but most importantly it appears, was black power in the ballot boxes (Calloway, 1965:14). Tucker said (in an interview with the author, March 7, 1967, St. Louis, Mo.) one of the Negro politicians related: "All the civic improvements did not put bread in our bellies." The black wards which he had formerly carried switched their allegiances to the man who gave financial support to the leaders of the Jefferson Bank and Trust Company sit-in.

In the subsequent mayoralty campaign between Cervantes and Maurice Zumwalt, lackluster Republican, the former rode amongst the escutchcons of Civic Progress. Banfield (1965:127) predicted that Cervantes would win the support of the "newspaper wards" by playing the role of an elected city manager.

Banfield points to a more politically active corporate elite than acknowledged by Greer (1963), or Bollens and Schmandt (1965). Banfield (1965:121) identifies good government with upper and middle classes and says lower classes are indifferent and/or hostile. He argues that St. Louis should be among the worst governed cities because of its preponderance of low income inhabitants. Although an "unreformed" political system with partisan elections, the corruption of a generation ago, he says, is gone. St. Louis is better than she should be because of competing forces. There is the "good government group" made up of the Mayor's Office, and the group which consists of county offices and the board of aldermen which he refers to as the "bad government group." "Neither side is strong enough to put the other out of business, and each needs the other to survive. The Mayor's Office needs help in the machine-controlled wards on election day. The County Office group needs campaign contributions from downtown businessmen" (Banfield, 1965:125). Of course there is the patronage, about which Banfield comments, that is apparently considerable through the county offices. Banfield (1965:126) says: "The mutuality of interests between 'delivery' and 'newspaper' wards results in the election of 'good government'-minded mayors."

Civic Progress members work with the Mayor's Office on bond issues and community projects. Commonly mentioned are: J. Wesley McAfee (Union Electric), Harry F. Harrington (Boatmen's National Bank), and Preston Estep (Bank of St. Louis). Others like Alfred Fleishmann (public relations executive) act for Civic Progress on governmental matters. For example, Fleishmann in 1967 had training sessions for city hall employees on how to be good public servants. Also, the Governmental Research Institute, Hospital Planning Commission, and Health and Welfare Council assist the Mayor's Office, and these agencies are all controlled through

Civic Progress members. In the development of the Model City Agency proposal, staff were loaned by Union Electric and McDonnell Douglas.

Former Mayor Tucker (interview with the author, March 7, 1967) said, "The Mayor is not that powerful—people think he is." Besides being beholden to business elites, the Mayor's Office is curbed by the personnel department which administers the civil service system. Salary levels and job specifications are oftentimes not competitive with private business. Moreover, patronage is limited, which is particularly disadvantageous for blacks who sometimes do not score as well on standardized tests given to determine qualification for civil service positions. Consequently, federal funds which might be controlled directly by the Mayor's Office, are allocated to private agencies not hamstrung by civil service stipulations. Another pattern to circumvent civil service is for city agencies to contract with private organizations, using federal funds, such as was done with Robert Gladstone and Associates for economic studies (*Technical Report on Housing Policy Guidelines for St. Louis, Missouri,* St. Louis Development Program. Washington, D.C., November, 1968) regarding the St. Louis Community Renewal Program. This can weaken the mayor's power because outside technical expertise which is part of the American corporate framework generates basic facts which public employees presumably lack the competence to develop. Hence, the Mayor's Office generally is cut off from internal expertise that comes from advanced degrees, with which business and the nonprofit corporations are so favored, and likewise constrained at the other extreme by having few patronage jobs for the less literate but nonetheless politically potent.

Business leadership in city government has not been plentiful. Board members' activity in voluntary agencies traditionally has been characterized as a field of service for influentials with money. Boards of public agencies usually have less influence and members of lesser stature. With the present city administration, influentials embrace it cautiously because they fear Mayor Cervantes may embarrass them. However, they cooperate to safeguard holdings and fulfill roles as Establishment stewards. This reticence was absent with former Mayor Tucker whom they respected and more fully supported.

BRICK AND MORTAR

At the end of 1966, the *Globe-Democrat* reported that in eight years time $450 million had been invested in downtown redevelopment. It is a story basically of Civic Progress, the Mayor's Office, Missouri Redevelopment Law, the Federal Government, and general obligation bonds. Spin-offs of Civic Progress: Downtown, Inc. and the Civic Center Redevelopment Corp. played important roles in building the image, applying political pressure, and securing financial backing for development of Busch Memorial Stadium, parking lots for the stadium, and Mansion House with its apartments and offices. Individual companies benefitted: Anheuser-Busch got the jewel of the downtown: Busch Stadium; Mercantile Trust owns part of the $52 million Mansion House complex; both Pet Milk and Laclede Gas will have new buildings in the central business district. Clarence Turley, downtown realtor, and member of Civic Progress promoted the Mayor's Spanish Pavilion. Preston Estep, Bank of St. Louis, chaired the committees for bond issues so that local public funds could be matched for federal dollars.

According to the Chamber of Commerce's report, the idea of the stadium was presented to the Chamber in 1958 by Charles Farris, executive director of the St. Louis Land Clearance and Redevelopment Corporation (*St. Louis Commerce,* "The Dream that Became a Reality." May, 1966:7). Hickok, president of the First National Bank, headed the committee to look into the matter. General Sverdrup's firm volunteered to study its feasibility. The committee recommended that a limited profit corporation be established under the Missouri Urban Redevelopment Corporations Act.

Three incorporators were appointed by the Chamber's executive committee: Hickok, Estep, and Sidney Maestre (Chairman of the Board of Mercantile Trust Co.). These men created the Civic Center Redevelopment Corp.

Based on the redevelopment law, tax relief is given for twenty-five years: for the first ten years, the tax is on the land at its predevelopment value; the last fifteen years, assessments are based on the land in addition to 50 percent of the value of improvements. However, provisions of the law were circumvented by leasing the land from the Land Clearance agency (which pays no taxes) for thirty-one years, the length of the loan period, and therefore, evading tax payment. After the thirty-one years, Civic Center then would receive the twenty-five years of tax relief as provided by the Redevelopment Act.

What were the interorganizational lines of influence? The Mayor's Office has control over who will be the executive director of the Land Clearance agency since he appoints the committees who choose the director. Tucker was mayor when Farris was appointed. The incorporators either are or were, at one time, Civil Progress members. Moreover, the executive committee of the Chamber is well represented with Civic Progress members.

The Civic Center group took title to the thirty-four block area in downtown St. Louis under authority of a city ordinance approved by the board of aldermen in March of 1961 which was a tripartite agreement among the city, Land Clearance, and the Civic Center agency. Thus these civic aristocrats embarked upon a tax free $89 million project.

The area was described thusly (Civic Center brochure, n.d.):

> Where this beautiful Civic Center Busch Memorial Stadium now stands, was an eyesore consisting of vacant lots; empty, rundown, ramshackle buildings; a few industrial plants; some warehouses, and a number of parking lots. Those who knew most about this area and had intimate acquaintance with it were the police, fire marshals, and building and health inspectors. The only citizens who came in this area were either those who worked or had business there, or a few unfortunate human derelicts who had no place else to go.

Of course, this was choice land in the central business district. The use of law to take land and utilize it tax free adds credibility to the notion that the "rich get richer" while they engage in civic undertakings. John Ise, formerly economics professor at the University of Kansas, liked to say: "Thou shall not steal—in small amounts."

Through the efforts of the Mayor's Office and Civic Progress a $6 million bond issue was passed by city voters to foot the bill for streets and lighting in the Civic Center area. Since this was being done for Anheuser-Busch who owns the St. Louis Cardinals, it was not surprising that August A. Busch, Jr. purchased $5 million in securities for the project. About fifteen million dollars was put up by more than three hundred shareholders, but the major mortgage financing was done by the Equitable Life Assurance Society, about $33.5 million.

The Board of Directors of CCRC were: August A. Busch, Jr., President, Anheuser-Busch; David R. Calhoun, Jr., President, St. Louis Union Trust; Edwin M. Clark, past President, Southwestern Bell; Joseph P. Clark, President, St. Louis Labor Council AFL-CIO; H. Reid Derrick, President, Laclede Gas; Preston Estep, President, Transit Casualty, and Chairman of the Board, Bank of St. Louis; John Fox, President, Mercantile Trust; Harry F. Harrington, Chairman of the Board, Boatmen's Bank; James P. Hickok, Chairman of the Board, First National Bank; Morton D. May, President, May Department Stores; J. W. McAfee, Chairman of the Board, Union Electric; Raymond E. Rowland, Chairman of the Board, Ralston Purina, and Charles A. Thomas, Chairman of the Board, Monsanto (Civic Center brochure, n.d.). The only one not in Civic Progress at one time or another was Clark, the union man. The Advisory Board of twenty-one men had mostly either active or emeritus members of Civic Progress.

SPANISH PAVILION

It started back in 1965 when Cervantes persuaded the Spanish government to give its world's fair exhibit to St. Louis. The proposition appeared attractive since the Spanish Pavilion had cost about $10 million and St. Louis could have it for the costs of dismantling, packing and shipping. The price tag was $700,000 to get it moving toward St. Louis. Before a drive in December 1966 began to raise $2 million to cover costs of the venture, $1,200,000 had already been contributed. The Mayor's Spanish blood ran hot over the prospects.

However, Civic Progress influentials were disturbed by the Mayor's rashness. Before they would agree to have a campaign to raise $2 million, he had to agree to let them handle the money and negotiate the business end of it. For one thing, they insisted on a feasibility study and were assured by Economic Research Associates that the project appeared viable, and in the interest of all St. Louisans. The following (Swayze, 1967) gives some indication of the distrust of the Mayor:

> The inspiration and the initiative for bringing the "Jewel of the New York World's Fair" to St. Louis came from Mayor Alfonso J. Cervantes.
>
> But if the inner council of the business community was stirred by the Mayor's imagination, it remained unmoved for the most part by his approach.
>
> Some felt that the Mayor had entered into the project without adequate examination of the economics. They felt also—not in a retributive sense but simply as a matter of practical business—that no such project would go here without the active support of key business leaders. And without a careful examination of the economics, no support would be given.
>
> Another factor was also involved, the greater confidence St. Louis business leaders would have in the success of the pavilion if it was under their effective control.
>
> In a round of conversations late last year, the practical details of the project were worked out, and in the words of one St. Louis businessman, "Everyone is happy now, happy with the arrangements, and happy with the prospect of success."
>
> Most of the officers and directors of the Spanish International Pavilion Foundation are officers and directors of the Civic Center Redevelopment Corp. developers of the massive $50,000,000 stadium complex, in which the pavilion will occupy space under a ground lease.
>
> The other directors are community leaders of equal prominence. The full list of officers and board members reads like a roll call at a meeting of Civic Progress, Inc.

The Spanish International Pavilion Foundation let the bids and the lowest was for $2,860,000. Preston Estep announced at the same time that $1.5 million would be needed to finish the interior of the Pavilion. Based on what had already been spent, plus anticipated expenditures, the project was estimated to cost more than $5 million.

Cervantes could relax now, but earlier, business leaders had threatened to walk out on the whole plan:

> Top business leaders, who once were dubious and even openly opposed to the project, now are driving hard to raise the money needed to complete the job.
>
> The change in attitude is due to a number of factors, chiefly: One: A willingness on the part of the Mayor to divorce all present and future operations of the Pavilion from any political or "City Hall" influence. The entire project is now in control of the same business leadership that successfully pushed through construction of the Busch Stadium complex and is responsible for most of the $500,000,000 downtown riverfront development.
>
> Two: The offer last spring by the Carondelet Savings & Loan Association that it would provide $2,500,000 a self-liquidating 30-year loan, if $2,000,000 in additional private donations can be collected. . . .
>
> Business leaders such as James P. Hickok, board chairman of the First National Bank of St. Louis; John L. Wilson, board chairman of UMC Industries, Inc.; Clarence M. Turley, downtown realtor, and John Fox, board chairman of Mercantile Trust Co. decided to go all out for the needed additional $2,000,000. . . .
>
> And this has come about only because hard-headed business management is running the show.
>
> Mayor Cervantes has never had a moment's doubt that the project will be a great success, both culturally and financially. . . .
>
> However, there has been criticism over the speed with which he made the commitment; unhappiness over a Mayor's Office acting as a fund raiser and general criticism over the inaccurate estimates of the costs of the entire project.
>
> . . . the initial success of the Gateway Arch (where visitors already are waiting in line for hours to ride the train system even without the added lure of the magnificent museum), plus the pulling power of beautiful Busch Memorial Stadium has not been lost on the hard-headed business community.
>
> "It has been estimated that if the Spanish Pavilion encourages tourists just to spend one more day in St. Louis, the city will receive more than $55,000,000 in extra spending annually," said Mr. Wilson, who with Mr. Turley is handling the $2,000,000 fund drive . . . (Schafers, 1967).

The Pavilion will have restaurants specializing in Spanish, French, and Italian delicacies. It is expected that the Pavilion will have a number of international trade promotion exhibitions yearly.

Church leaders in St. Louis legitimated the Pavilion as being worthwhile for uplifting the people and producing revenues for the city.

McDonnell Douglas Corp. contributed $150,000. Half came from the McDonnell Personnel Charity Trust and the other half from the McDonnell foundation ("McDonnell Gives Pavilion $150,000," *St. Louis Post-Dispatch,* August 15, 1967).

The gift from the personnel fund was opposed by some employees who questioned the Pavilion's constituting a charitable undertaking ("McDonnell Replies to Fund Critics," *St. Louis Globe-Democrat,* August 17, 1967). The announcement of the gift was made first to the media and then to the employees. Notices posted at McDonnell Douglas explained that a three-man board elected by the employees decided the Pavilion represented a suitable charitable purpose. The fund's bylaws state money can be contributed for capital funds of community institutions.

A picture of the Mayor's son, Brett Cervantes, showed him making a contribution to the Charity Well in the lobby of the Greyhound Bus Terminal. "The well is part of a nation-wide charity program of Greyhound Corp. Each month the mayor will designate a charity to receive proceeds for 30 days. The International Spanish Pavilion Foundation, seeking funds to reconstruct the Pavilion in St. Louis, is the first group to benefit from the well" ("Mayor's Son Makes Gift to Charity Well," *St. Louis Post-Dispatch,* August 18, 1967).

TAX FREE STATUS CHALLENGED

Mr. and Mrs. Elmer G. Doernhoefer formerly owned two tracts of land where the Riverfront Inn was being built in 1967. Through condemnation proceedings they lost their property to the St. Louis Land Clearance agency which accumulated land for the Civic Center. The Doernhoefers appealed this action on the grounds that the Land Clearance agency was not the real purchaser, that in fact it was the Civic Center who was benefitting illegally from a tax free status. The Doernhoefers settled out of court before a decision could be made as to whether the arrangement with CCRC was in violation of Missouri law ("Stadium Area Land Status Unresolved," *St. Louis Post-Dispatch,* November 5, 1967).

An editorial in the *St. Louis Post-Dispatch* ("Tax-Free at the Stadium," November 9, 1967) said:

A challenge to the constitutionality of a device freeing the Downtown Stadium, its parking garages and the new motel of all real estate taxes for up to 31 years has been dismissed, which is a pity; the question ought to have been resolved, and still should be. But constitutional or not, the tax avoidance scheme surely represents questionable public policy, going as it does beyond the letter of the Missouri Urban Redevelopment Law which provides only partial tax abatement and then only for 25 years.

As the situation stands, Civic Center could avoid any real estate taxes for 31 years then begin the 25 year period of partial tax abatement under the Urban Redevelopment Law, giving the project 56 years of total or partial tax relief. Even if this particular contract is legal, we doubt the wisdom of permitting others to take advantage of the same route.

For its part, the Civic Center Corp. ought to begin making voluntary payments in lieu of taxes as soon as it is financially able to do so, probably when the Riverfront Inn becomes operable. Certainly payments in lieu of taxes should come ahead of dividends to the Civic minded individuals and groups that hold equities in the corporation.

Republican Alderman Lee Weber proposed investigating the tax free status of the Civic Center. This includes in addition to the Stadium, the Spanish Pavilion and an office building which is to be built. The tax loss is estimated to be running at about $2 million a year.

Careful analysis might demonstrate the special tax treatment devised for the Stadium ought to be modified if legally possible, and the question of how best to modify it might properly occupy an aldermanic inquiry. For instance, Civic Center has agreed to make payments in lieu of taxes if it meets its mortgage debt. But what of the years it can't pay? ("For Answers on the Stadium," *St. Louis Post-Dispatch*, November 16, 1967).

The City Counselor had never ruled on the tax status. No taxes had been paid for 1964, 1965, 1966, and assessment cards for 1967 were marked "exempt" ("M'Guire's Opinion Sought on Stadium's Tax Status," *St. Louis Post-Dispatch,* November 28, 1967).

In December the following article ("City to Get $150,000 in Stadium Funds," *St. Louis Globe-Democrat,* December 18, 1967) appeared in the *Globe*:

"The city will shortly receive about $150,000 from the Civic Center Redevelopment Corp. in lieu of taxes for 1967 from profits on the operation of the Busch Memorial Stadium complex." The executive director of CCRC said payment was not being made because they were under pressure; it was something they were planning to do even before the tax matter came up.

Later that month, the following article ("Blight Study Ordered," *St. Louis Post-Dispatch,* January 19, 1968) stated: "An over-all study of blighting in the downtown area was ordered last night by the City Plan Commission. The study was requested by Downtown St. Louis, Inc.

"The request comes in the wake of a series of proposals to declare certain blocks blighted under a state redevelopment law, which gives developers tax relief for up to 25 years."

Not shown in Figure 5 are nonpreferred public and private agencies peripheral to the political organization and governance even though they are part of the Establishment. Of course, many other locally based public and private agencies are either interconnected with or detached from agencies surfaced for this analysis. Besides those indicated, federal and state agencies give services and income to people in the city. Reality is indeed complex, but it is not the purpose of this study to recount the multitudinous organizational actors in the community's social welfare systems. They are too numerous and can be more appropriately enumerated in a directory.

Rather, this is a model of the St. Louis Establishment, gatekeepers of a political order based on substantial inequities in the allocations of material resources. Their staffs are experts in community organization, individual and group therapies. By and large this cadre is committed to change which will improve the quality of living through a reasoned consensus instead of protest and conflict. First and foremost, they are brokers of corporate elites whether they work in the private or public agencies. At times, they do little more than defend targets of protest (Lipsky, 1967:22–27). Basically, they symbolically assure and tangibly reward so that the established order will change without perishing. Their expertise is used to design individual and group therapies to facilitate adjustment and maximize opportunities. They are also stewards of income strategies based upon finding jobs for the disadvantaged in training programs, subsidized employment, and in the workaday world even though these jobs are often at marginal wages; but such are the token material satisfactions to which Lipsky (1967:24) refers.

Through the agencies in this model, major decisions are made in nondemocratic and nonpluralistic ways about the organization of social welfare, the amount and nature of community services. The justification for exposing these agencies to public scrutiny lies in their being extraordinarily influential both economically and politically. If there were other agencies which held such major strategic positions in policy making, then it would not be justified to exclude them from the model. However, none but these appear to dominate the political process in which the organization of welfare occurs.

In sum, rulership is by an interrelated establishment of corporate and public influentials assisted with a cadre of policy scientists and "diplomats" who manage the destinies of thousands of people in terms of a welfare model of social problems which defines deviant behavior in therapeutic instead of political terms (Horowitz and Liebowitz, 1968:280).

CONCLUSIONS

Figure 5 captures the spirit of the chapter in depicting the political organization of social welfare centering in Civic Progress and not in the Mayor's Office as one might suppose.

Dependence of city officials on business influentials is demonstrated through the latter's control of the Civic Center and Spanish Pavilion. In other parts of the book the role of business influentials in bond issue campaigns is documented. Both the mayor and other elected officials need the talents and money of finpols in running for office. Tax duplicates rise and fall based upon actions of large corporations.

The payoff for business is that public servants are brokers for expenditure of public monies and use of city cadres, but also, most importantly, in terms of profits. Such reciprocities augment the legitimacy and prestige of civic giants. Private government, not public government, is viewed as the way to get things done.

Chapter 9

Mayor's Office:

Social Welfare

Social welfare pertains specifically to health, crime, social services, employment and the like. It is distinguished substantively from "brick and mortar" social policy. Social welfare concerns human beings, their life styles and life chances, related to the physical side of city development, but inadequately because of the broken image (Matson, 1966). Social welfare is the excrescence resulting from failure to apply a public ideology; such services, then, make private ideology work. This is not to say all services are like this, because health care for instance, is needed by most at one time or another; however, health care as a social welfare problem means people unable to pay for medical care.

This chapter, as the previous one, describes the interpenetration of the Mayor's Office by an interorganization of influential agencies: semipublic and private in the sense that the latter are more or less independent of public regulation by either funds or statute. Semipublic or semiprivate agencies are de facto public but de jure private. The Human Development Corporation, financed by OEO for the War on Poverty, was voluntarily constituted by law; it is public in nature because it could not continue without public funds and moreover, its basic operations are stipulated by the Economic Opportunity Act and the concomitant administrative interpretations. The Civic Center Redevelopment Corp. achieved a public-like status because of a city ordinance setting it up, its existence being premised on the Missouri Redevelopment Corporation Act, and its tax free status deriving from having leased property from the Land Clearance Authority.

Nevertheless, it is de jure private since its incorporators have the power to dissolve it. Actually, the Civic Center is not dependent on public funds and utilizes these on its own terms rather than having them dictated by legislators or laws.

EWGCC

The East-West Gateway Coordinating Council is also an example of a de jure voluntary but de facto statutory agency. In order for the St. Louis area to qualify for highway funds under the U.S. Highway Act of 1962, it was necessary to form a comprehensive and continuing planning body (Luna, 1967). The EWGCC is this agency and was instituted in 1965. Its planning area having 3,567 square miles, 439 taxing bodies and a population of 2.5 million, includes the City of St. Louis, St. Louis, St. Charles and Jefferson Counties in Missouri, and three counties in Illinois. The Council is a council of governments with thirteen voting members of whom six are chief elective officers from the Missouri side, and six are from the Illinois side of the Mississippi River. The other voting member is the chairman of the Bi-State Development Agency. A majority is required from both sides of the river for approval of proposals. Its source of support is from local and federal funds. Local political entities contribute 5¢ per capita, and the federal government 10¢. In order to qualify for federal funds, at least 90 percent of the population must be included in the area encompassed by the Council; EWGCC has about 97 percent. The Council is under the supervision of the Housing and Urban Development Department.

The scope of EWGCC was extended beyond highway planning with the passage of the Demonstration Cities Act which gave authority to such councils to review all applications for federal funds for hospitals, libraries, parks, recreation, and other such related facilities (Sutin, 1966:1–2). One of its first jobs concerned bi-state air pollution control. This new power is to deal with the pluralism of governments in metropolitan areas.

The EWGCC would seem to be independent of Civic Progress, Inc. However, such is not the case. It is subject to pressure from the large corporations. Also, because only one vote is given to each government regardless of size, the problem concerning its representativeness is a source of friction. Such divisiveness makes it easier for streamlined organizations such as Civic Progress to dominate. They have only to divide the governments to block legislation or policy changes.

The mayor of St. Louis has one vote in EWGCC, the same as municipalities much smaller. Not only is representation not according to population, the mayor must submit to planning decisions which encompass a wider area than would be circumscribed by the narrow interests of St. Louis itself. This creates an agency which in effect reduces the mayor's powers in the city, and Mayor Cervantes considers this phenomenon to be onerous.

HDC

The Human Development Corporation is another example of a private agency with a statutory basis. In 1962, the Juvenile Delinquency Control Act was passed by Congress in response to the growing pressure for Negro rights. The President's Committee on Juvenile Delinquency, given responsibility for implementing the Act, premised its program upon a battery of services designed to eliminate blocked access to opportunity. Remove the barriers and Negroes could move up in socioeconomic status. Provide services, open up job oportunities, and the problems of delinquency, teen-age unemployment and demonstrations would be curbed, was their reasoning. Existing private and public, social and health services were attacked as being insufficient and inaccessible to people needing them most.

Although HDC had its start through the Mayor's Office, being a private agency gives it autonomy to operate in ways similar to a business corporation but at the same time having access to public funds. As noted before, although the Mayor of St. Louis and officials of public and voluntary organizations obtained funds for what was called the Juvenile Delinquency Control Project, President Kennedy was assassinated before the planning stage could be put into action locally.[10] The new President shortly thereafter declared war on poverty.

In anticipation of new legislation on poverty, a program to deal with problems other than delinquency was outlined. The name of the project was changed to the St. Louis Human Development Planning Project in the Fall of 1963. By December of that year, a voluntary nonprofit corporation was formed.

From the start, the Establishment's strategy was to partially cooperate with the Human Development Corporation. As mentioned in Chapter 4, the finpolity was at bay with the generally deteriorating community situation, and the criticism being lodged against its community services. The *St. Louis Globe-Democrat* supported the Establishment by creating a bad image of HDC through emphasizing the waste, and high salaries of its employees. Only a few UF agencies became involved, as was also true elsewhere in the United States.

Wayne Vasey, Dean of the School of Social Work at Washington University, General Manager of HDC temporarily for a year, submitted his resignation in 1965. MacDonald (head of HWC), and others were concerned that Mayor Cervantes might gain control of HDC and exploit it for his own political ends through appointing a political crony. In a memo ("Poverty Planning in St. Louis," HWC, July 19, 1965) prepared for Council board discussion, the following was stated:

> In the last week of June, the HDC Executive Committee proposed to the Council President that HDC be merged with the Council or that the latter, by contractual arrangement, take over certain planning functions in the poverty program. Staffs of both organizations were directed to prepare an agreement to be considered by both boards.

The merger with HDC did not happen; MacDonald backed off from the responsibility. He already had HPC to reconcile and a confrontation with the Metropolitan Youth Commission was brewing. Soon afterwards, Sam Bernstein, Director of the Jewish Employment Vocational Service and successful in obtaining thousands of dollars in federal funds for that agency, was chosen to be Vasey's replacement as General Manager of HDC. Tested in the UF agency; a known quantity to the civic influentials and a social worker by training, Bernstein could be expected to serve the Establishment.

Attempts to democratize the Human Development Corporation were abortive. Newly appointed representatives of the poor tried to change the bylaws to permit their direct appointment to the board of HDC. They objected to the system calling for the Citizens Advisory Council to nominate; the Mayor and Supervisor to select from those names. (The Mayor chooses three from six and the Supervisor two from four names nominated.) Cohn, representing the Supervisor, asked that the latter be able to appoint one of the ten nonpoor board members now named by the Mayor of St. Louis. Father Lucius Cervantes (the Mayor's brother) commented that Sergeant Shriver, leader of OEO, was no longer antagonistic to mayors and was considering permitting them to be voting members of antipoverty agencies. Thus the Board, Father Cervantes suggested, should hold off in revising the bylaws so that changes regarding representation and voting rights could be made at the same time (Shepherd, 1967).

Protest leaders have found employment at the Human Development Corporation. Staff member Norman Seay, formerly a Democratic committeeman in the Twenty-sixth Ward, was in the Jefferson Bank & Trust sit-in. He received a $50-a-month pay increase after having resigned as CORE's chairman of the boycott of bus service (Jacobs, 1966). CORE conducted "freedom rides" to protest selling the Consolidated Service Car Co. These service cars were oversized taxis that made scheduled runs through the black belt. Consolidated (Mayor Cervantes was formerly part owner) sold out to Bi-State Transit Company for $625,000, and some

Negroes were displeased to have the poor black man's transportation forced out of business to make way for the white man's busses. Seay had to decide either to give up his CORE position or resign from HDC. Another vice-chairman of CORE also involved in the freedom rides was promoted to neighborhood developer in the West End Gateway Center ("Anti-poverty Unit Elects 32 New Members," *St. Louis Post-Dispatch,* November 15, 1966). One militant civil rights leader became a teacher in a program founded by HDC. Percy Green, frequently arrested for activities in connection with ACTION, a protest group, was employed at $7,500 a year ("Percy Green Teaches in Poverty War Program," *St. Louis Globe-Democrat,* January 7–8, 1967). Since being hired by the Jewish Employment and Vocational Service, a contracting agency with HDC, he has participated in no illegal demonstrations according to a report from his supervisor.

Critics of the Human Development Corporation, which is frequently a target group itself, have often been silenced. The local head of the Urban League had been openly critical of HDC, but when his agency received substantial funds from HDC, his public stance changed. Another critic of HDC, former city editor of the *St. Louis Argus,* was appointed executive director of the St. Louis Small Business Development Center, Inc. ("Editor to Assume Post of Small Firm Agency," *St. Louis Globe-Democrat,* May 23, 1967). He had opposed high salaries paid HDC workers.

The handling of protest by HDC is illustrated by activities of the Pruitt-Igoe Neighborhood Corporation, which was organized by Robert Wintersmith, coordinator of the Urban League's HDC financed antipoverty station.

The following happened: (1) Wintersmith in July obtained $32,000 for a 10-week program to hire 76 people from Pruitt-Igoe to work as organizers—"getters" and "givers" of information; (2) Tenants were invited to meetings to talk about problems in the project; (3) The tenants made a study and wrote a report which featured their complaints about the project and its management with immediate demands to either correct the conditions or put the project under federal control;[11] (4) Top management of the Housing Authority resigned; (5) Early in January, because little had been done and there were new complaints, the tenants called a rent strike for March 1 unless their demands were met (this was made public January 11 during the time the board of aldermen was debating giving HDC the $25,000 salary money); (6) The *Post-Dispatch,* usually the downtrodden's champion, opposed the rent strike in spite of the charges that tenants were "freezing to death"; the reasoning offered by the *Post* was the rent strike would be self-defeating since the Housing Authority depends upon the tenants for its operating income ("The Rent Strike," an editorial. *St. Louis Post-Dispatch,* January 1, 1967); (7) The executive director of the Urban League tried unsuccessfully on January 12 to meet with tenants about differences on the rent strike; (8) The Aldermanic Housing and Land Clearance Committee attempted to arbitrate the dispute January 20; new management of the Authority asserted that $1,200,000 to $2,000,000 in improvement programs had been contracted, but that more cooperation was needed by the tenants; moreover, Pruitt-Igoe was the only one of the projects causing this much trouble; (9) Toward the end of January, the Pruitt-Igoe Neighborhood Corporation released a letter to the Mayor asking for representation on the board of the St. Louis Housing Authority; (10) January 24, tenants warned the aldermanic committee of a new Watts at Pruitt-Igoe; this was a "13-year grip," one said; there were sharp clashes between aldermen, housing authority people and tenants at this meeting concerning broken promises; (11) A House Committee in Jefferson City approved a resolution to investigate Pruitt-Igoe; (12) Federal authorities moved in on Pruitt-Igoe in January —met with Mayor Cervantes, housing authority officials, HDC, and others concerned; (13) On the following day Eugene Porter, chairman and usual spokesman during the revolt of the Pruitt-Igoe Neighborhood Corporation, said the threat of a strike was eased because of the meeting with the federal officials; that the differences were not great between the housing officials and the tenants; (14) By the 24th of February the strike had been postponed indefinitely in order to give time to make improvements; HUD had asked for a two-month postponement, but Porter said that they would hold off as long as satisfactory progress was being made on changes; he called it a victory for both sides in that the Housing Authority understands the situation much better; federal people had promised the work would be done; (15) That same weekend it was announced that Mrs. Floyd, vice-chairman of the Pruitt-Igoe Corporation and articulate opponent of the authority, had been invited by HEW to Washington, D.C. to serve as a consultant on a three-day briefing session.

The account indicates how private and public agencies and their influentials interacted in ways to avert or abort revolt. Professor Rainwater (in consultation with the author, November 15, 1967, St. Louis, Mo.) states Wintersmith was bluffing because he knew that he did not have a united community. The *Post-Dispatch* served the logic of the business interests well by rejecting the strike for economic reasons.

DELINQUENCY PLANNING

The Metropolitan Youth Commission represents another type of agency relationship with the Mayor's Office and the Establishment. Once again, the dependence of the mayor on the private agencies is apparent. MYC had been critical of the Health and Welfare Council's lack of action on issues considered important by the Commission. Also, the Commission, unsuccessfully fitting into HDC's operations, expected HWC to be more supportive. Developments of a bizarre nature took place during the summer of 1965, presenting MacDonald with another occasion to augment his own agency.

During the summer, while the director of MYC was in Europe, Father Lucius Cervantes[12] investigated this agency. After poking through files and research reports as well as interviewing executives and board members, he concluded MYC was not unifying delinquency services and found its research irrelevant. Subsequently, Mayor Cervantes decided to abolish the Commission which had been set up in 1956 by the City of St. Louis and St. Louis County ordinances and financed jointly.

Actually, the Metropolitan Youth Commission was an outgrowth of a Council study of the 1950s. Concern about increasing juvenile arrests led to a curfew law and recommendations for greater coordination of community social welfare services and law enforcement authorities. The Health and Welfare Council in a study (Archives of the Health and Welfare Council of Metropolitan St. Louis) made in 1955, recommended that a youth commission be established. Former Mayor Tucker some years later (in an interview with the author, March 7, 1967), cited the organizing of MYC as part of the government of the City of St. Louis and St. Louis County, separate from HWC, as a failure of the latter to really do anything about consolidating various health and welfare services.

In October 1965, the Health and Welfare Council's Executive Committee ("Report to Board of Directors," HWC, 7, October 26, 1965) proposed that:

> 1. The Commission be retained by the City and County and brought up to full strength. 2. The Commission contract with HWC to provide staff to work in the youth delinquency area, and 3. The Commission and Council integrate delinquency planning with other HWC programs. Letters of intent would be exchanged and a formal contract agreed to by all parties concerned.

After six months of operations with HWC staff, a program report was made to the Mayor and County Supervisor, contractors with HWC for providing staff services for the Commission. Father Cervantes objected to the population studies done jointly by the Commission and Council as being more appropriate for the latter or some other agency to do. One report to which he referred estimated the city's population to be 50,000 less in 1965 than in 1960 ("Population by Census District St. Louis City," HWC and MYC, January, 1967). The Mayor is defensive about studies showing that the city's population has declined sharply since the last census.[13] The Mayor wants to show the city to be attractive for industrial development, and views other parts of the metropolitan area, particularly the county, as rivals.

Subsequently, MYC was informed the Mayor lacked funds to continue the city's part of the contract. Poelker and Donald Gunn (President of the Board of Aldermen) intervened and the Mayor agreed to continuance of the contract if the Aldermen raised taxes.

During the months of uncertainty about MYC's financial future, Father Cervantes asked for assistance of the Commission in writing certain parts of the Model City proposal pertaining to delinquency programs.

Delinquency control was reconceptualized from grassroots citizen activity into a welfare model. While MYC formerly organized citizen groups for action on youth problems, this was discontinued with the transfer of this agency to the Health and Welfare Council. The previous activities of the commission stimulated protest against the institutional management of delinquency behavior, but had been unable to channel this protest in terms of a consensual model. There was a failure to unify services. The new program of the Commission emphasizes the evaluation of different strategies for control of juvenile misconduct. The focus is on situational manipulation of the environment such as recreation programs, leadership training through camping experiences, and experimental attempts to modify behavior by providing rewards for desired behavior (Archives of the Metropolitan Youth Commission, and *HWC 56th Annual Report,* Health and Welfare Council of Metropolitan St. Louis, 1967). A Junior Aid Program was developed in conjunction with the St. Louis Police Department and YMCA to develop more positive attitudes toward the police ("36 Teen-Age Boys Join Police in 10-week Junior Aid Program," *St. Louis Globe-Democrat,* June 30, 1967). Therefore, instead of encouraging citizen action against institutional policies or societal conditions, the model is therapeutic.

HEALTH CARE

The following transactions illustrate the use of the Mayor's Office by the private sector to secure health care objectives.

The Hospital Planning Commission's financial future was unsettled, also, because the Mayor's Office had not made an allocation for its support in 1967. A similar application to the County Supervisor's Office had not been approved either. Money from City and County governments is needed (in addition to the allocation from the UF) to expand HPC's program as required by the United States Public Health Service. In order to continue receiving Hill-Harris funds (P.L. 88–443), the following recommendations (United Fund of Greater St. Louis, Inc. Agency Budget Request for 1967) must be accomplished:

(1) The HPC should assume a leadership role in planning long-term care facilities within its area. Present work includes only an annual analysis of facilities and review of a few projects upon request. (2) It is essential that the HPC adopt a detailed set of criteria for planning and review. This means considerably more specific criteria than used up to the present time and will require extensive work. (3) The HPC should enter into patient origin studies to enhance its knowledge of mobility of patients and determining areas of greatest need. (4) The six-county area must be broken down into smaller planning grids since projection of target needs on a larger basis is susceptible to attack because of lack of preciseness. (5) The present staff is inadequate for the area to be serviced and planned. The consultants recommend that at least three more professionals be added, but the Commission must take the initiative in securing local matching funds and enlarging the program. (6) If the Commission is to encourage long-range planning by each health facility, it must be in a position to assist each facility in relating its plan to community need and other neighboring health facilities.

These recommendations were the result of a site visit of USPHS staff, and a consultant who is the head of a hospital planning commission in another city. The letter informing HPC what it must do came not from the USPHS but from the private consultant.

A year previous to the aforementioned recommendations, the USPHS required the Commission to broaden its representation to include physicians and hospital administrators. Although there were mostly Civic Progress elites on the board then, several grinned and shook their heads.

Early in 1967, the HWC executive proposed that the Hospital Planning Commission merge with the Council if additional funding could not be obtained for the Commission. The board of the latter would not accept the proposal. John Poelker, two years before while President of HWC, could only get HPC's board to agree to a partial merger combining the executive directorships of the two agencies. Blue Cross and Hill-Burton consultants also were opposed on grounds the Commission might lose its identity and influence over hospitals.

Instead of total merger as proposed by the HWC-HPC executive, HPC's board recommended its financial plight be brought to the attention of Civic Progress, Inc. Mayor Cervantes, present at the meeting with Civic Progress, enthusiastically indicated the Commission would receive the money from the city whether the County Supervisor, Lawrence Roos, made an allocation or not. W. R. Persons, President of Civic Progress, assumed responsibility for talking with Supervisor Roos about St. Louis County's matching the city's allocation to the Hospital Planning Commission.

Related to the preceding point is the question of what will be the central agency for health planning in the St. Louis area. Executives of the Health and Welfare Council and the Hospital Planning Commission envision themselves in this role, but there are competing agencies. This question is a vital issue because of the Hill-Staggers Bill (P.L. 89–749) which compels comprehensive health planning for state and regional areas. The Office of State and Regional Planning and Community Development expected to be the statewide health planning agency, and had designated the East-West Gateway Coordinating Council as the central planning body for the St. Louis area.

Dr. Herbert Domke, head of the Health and Hospitals Division of the City of St. Louis is an advocate nationally, as well as locally, of comprehensive health planning. One of the conditions for his coming to St. Louis was for Mayor Cervantes to commit himself to this need in the St. Louis area. Neither Domke nor the Mayor had publicly stated HWC-HPC should be the agency for coordinating health care.

In a joint staff meeting with HWC-HPC-MYC and EWGCC, Eugene Moody, director of the latter organization, commented (Minutes of HWC Staff meeting, February 28, 1967): "I'm a negative empire builder." Although preferring to work through the existing agencies, he said that recent meetings with Housing and Urban Development Department officials point to their wanting more than a cursory review of projects proposing use of federal funds which include health and hospital facilities along with other affairs relating to the general welfare of the people. This would mean a need to expand his own staff. The HWC-HPC-MYC executive asked several times during the meeting if Moody would be agreeable to drawing up a proposal that HWC . . . "will work with EWGCC particularly in areas of its (HWC's) competence, e.g., hospitals, nursing homes, social services, etc. This would demonstrate to HUD that there is strong community backing to EWGCC locally."

The foregoing, concerning the Hospital Planning Commission and comprehensive health planning, is indicative of the political divisions based upon municipal and county sectionalism in the St. Louis area. Civic Progress, Inc. and the nonprofit, private corporations (e.g. HWC-HPC and EWGCC) bridge these parochialisms. Economic, and not political interests are generic in local politics; Civic Progress corporations overarch county boundaries. Federal funds and federal laws seem to undercut the autonomy of hospitals and private dominance in health planning. Political interests are being forced to act regionally if they want to continue to receive federal funds. However, the agency which emerges as ruler will likely have the power of the State at its command, and be more effective in imposing a rational and totalitarian model on the health field. In spite of constraints, the Hospital Planning Commission remains a self-perpetuating board dominated by Civic Progress influentials. Its linkages are more with Health, Education and Welfare at the federal level, and of course EWGCC is tied into the Housing and Urban Development Department. The struggle at the federal level for dominance has apparently descended to local areas.

Besides meeting with Mayor Cervantes, Domke and Dr. C. Howe Eller (head of St. Louis County Health Department), and others about health planning, HWC-HPC has obtained and interpreted facts on the Medicare legislation in relation to the utilization of Medicare in the State of Missouri, and made specific recommendations for the state legislature to adopt. But this was done in conjunction with key private and public executives.

A Medicare Task Force was organized by HWC-HPC with Howard F. Baer as chairman. Baer is a retired chairman of the board of A. S. Aloe Company. Various subcommittees were appointed—home care resources; personnel needs; nursing homes; and Title 19 (provides medical care for welfare recipients and others not on welfare but medically indigent) of the Medicare Act. The HWC-HPC and the Mayor's Office were instrumental in getting the state legislature to pass legislation enabling Missouri to be eligible to receive the federal money (*HWC 56th Annual Report,* Health and Welfare Council of Metropolitan St. Louis, 1967, p. 3).

Because Governor Hearnes had promised not to raise taxes, and due to alleged fiscal shortages, he did not want to implement Medicaid as quickly or as fully as local interests in St. Louis and Kansas City who joined forces to persuade him to alter his position. The advantages to the local areas are that voluntary contributions and tax funds for medical care of the needy can be reduced. However, the cost to the State of Missouri was estimated at $19.3 million (*Title XIX for Missouri,* Health and Welfare Council of Metropolitan St. Louis and Metropolitan Hospital Planning Commission, Medicare Task Force, October, 1966). Nevertheless, influential Democrats and Republicans alike backed Medicaid. Even W. B. McMillan ("Roster of the Right Wing in Illinois and Missouri," *Focus-Midwest,* 3, No. 6, 1964:14), a conservative, and vice-president of Hussman Refrigerator division of Pet Company, used testimony prepared by HWC-HPC staff. Frederic Peirce in Civic Progress backed Medicaid. General American Life Insurance, of which he is the chief executive officer, is a fiscal intermediary for Medicare.

Medicaid is a services strategy based upon the concept of public community welfare which has already been discussed in this study. It is advantageous to corporate elites to spread the cost of medical care for the needy over the populace. Even more importantly, it is a substitute for adequate income which would be a just way of meeting medical needs, but more expensive to the corporations.

MODEL CITIES

The Model City Agency in St. Louis, as in other cities, has been primarily generating and revising plans since 1966. With the change in administrations nationally, the role is yet to be clarified. Thus, the St. Louis Model City Agency cannot be judged in terms of what it has done, so much as it can, on what it said could be done. Although this agency started out structurally like the Human Development Corporation, subsequently it was established as a part of the city government under the administration of the Mayor's Office.

Members of Civic Progress, and lower echelon executives from Civic Progress companies assisted in the development of proposals broadcast throughout the community. The McDonnell Douglas Company loaned several of its executives in order to develop plans for a systems approach. An early proposal (prepared by the Model City Agency and HDC of Metropolitan St. Louis) was for a Metropolitan Housing Corporation, a private nonprofit organization, to find and finance nonghetto housing for Negroes. This was played down as the emphasis shifted to gilding the ghetto.

Community effort to establish a Model City Agency is illuminating from the standpoint of civic influentials participating in planning of a program which is to provide comprehensive services for rebuilding the slums of St. Louis and also because of the implications for containment of low income politics.

A steering committee was set up in the summer of 1966 to guide the development of an application for a planning grant which was to be obtained from the Housing and Urban Development Department under the Demonstration Cities Act (P.L. 89–754). This committee consisted of:

Honorable Alfonso J. Cervantes, Mayor, City of St. Louis
Arthur G. Baebler, Manager, Industrial Development Division, Union Electric Company
Jerome Berger, President, Laclede Town Company
Samuel Bernstein, General Manager, Human Development Corporation
A. Donald Bourgeois, Deputy General Manager, Human Development Corporation
Dr. Joseph P. Cosand, President, Junior College District
Dr. Herbert Domke, Director of Health and Hospitals, City of St. Louis
Irvin Dagen, Acting Executive Director, Land Clearance and Housing Authority

Charles L. Farris, President, Urban Programming Corporation of America

Dr. Robert Felix, Dean, St. Louis University Medical School

Alfred Fleishman, Fleishman-Hillard, Inc.

Mrs. Frankie Freeman, Associate Counsel, St. Louis Housing Authority

Harold Gibbons, President, Teamsters Joint Council No. 13

Dr. F. William Graham, Vice-President, Junior College District

George Hellmuth, Vice-President, Hellmuth, Obata and Kassabaum

Albert B. Hensley, Jr., Assistant Director of Development, St. Louis Housing Authority

Robert E. Hofmann, Analyst, Advanced Production Planning, McDonnell Aircraft Corporation

Robert B. Jones, Director of Planning, City Plan Commission

Ken Lange, City Plan Commission

Jake McCarthy, Editor, Teamsters Joint Council No. 13

Dan MacDonald, Executive Director, Health and Welfare Council of Metropolitan St. Louis

Harry L. McKee, Director of Long Range Product Planning, McDonnell Aircraft Corporation

Eugene Moody, Executive Director, East-West Gateway Coordinating Council

Guy Obata, President, Hellmuth, Obata and Kassabaum

Shelby Pruett, Architect, Hellmuth, Obata and Kassabaum

Dr. Miller B. Spangler, Director of Economic Research, St. Louis Regional Industrial Development Corporation

Kenneth S. F. Teasdale, Attorney at Law, Armstrong, Teasdale, Kramer and Vaughan

Herman M. Wagner, Planning Commissioner, St. Louis County

Although this committee had no Civic Progress members, it did have lower level executives from large corporations, and from strategic agencies on whose boards Civic Progress members serve. The focus on physical aspects is indicated by the committee's composition.

Fourteen subsystems in the following areas were proposed (*St. Louis City Demonstration Proposal,* September 30, 1966): (1) Housing; (2) Economic development; (3) Employment; (4) Education; (5) Health; (6) Community services; (7) Recreation and culture; (8) Civil Rights; (9) Transportation; (10) Antipoverty programs; (11) Utilities; (12) Communications; (13) Sanitation; and (14) Metropolitan St. Louis Planning. Various agencies were asked by the Mayor's Office, assisted by Father Cervantes in the Model City Agency effort, to write program proposals for the different subsystems. The Human Development Corporation was given the assignment of writing the over-all proposal for the planning funds. By the end of 1966, Donald Bourgeois, formerly deputy manager of the Human Development Corporation, was employed to direct the agency in the making. The staff of the Health and Welfare Council speculated that the Model City agency might supersede the Human Development Corporation.

One of the new approaches to involving low income people was to be a game called Trade-Off.

It may be played (and has been played) by professionals and nonprofessionals both separately and together. Community needs are stated in a situation prospectus. Example:

THE MODEL NEIGHBORHOOD

Population	9,500
Dwelling Units	3,800
Substandard Dwelling Units, Rehabilitable	500
Substandard Dwelling Units, Non-Rehabilitable	1,200
Elementary Schools Overcrowded and Obsolete	
(Elementary School Population)	2,000
Parks (1,000 people per acre)	9.5 acres

and so forth

Then the players are asked to build the best possible community. This may be done on a grid system laid out on a board using small wooden blocks representing physical elements. Perhaps the game should be played on a blocked-off street in the neighborhood, laying out the grid system with white lines on the street itself, and using large styrofoam blocks to denote physical improvements, meeting needs as stated in the situation prospectus. Each physical component may have a price tag or the entire community may be priced out component by component after the community is laid out. Example:

500 Rehabilitated Dwelling Units @ $4,500 =	$2,250,000
1,000 New Dwelling Units	@ $12,000 =	$12,000,000
New School	each	1,000,000

and so forth

Supposing, then, that the price tag for the entire community comes to 40 million dollars. We find now that 40 million dollars is not available. In the game, perhaps 25 million is available. Now the players are asked to assign points to each physical component selected. A yardstick may be established to speed the game. Example:

Each Rehabilitated Dwelling Unit = 10 points

The players are then asked to rebuild the community buying the most possible points per dollar for 25 million dollars. If there are competitive teams, the team buying the most points would win. Or a critique may be offered by spectators of the finished product.

"TRADE-OFF"

1. *Objective*
   (a) Understanding by planners of neighborhood values and goals
   (b) Understanding by players of differences in planning and decision making
   (c) Some understanding of cost benefit analysis
2. *Rules*
   (a) as above
   (b) anyone plays
   (c) can be competitive or otherwise
3. *Style*
   (a) Realistic
4. *Serendipity or Spin-off*
   (a) On-going communication
   (b) Planning involvement
   (c) Acceptance

These are some games. There are others. More games will be devised and discovered by planning teams and sub-cities themselves as the planning process continues. Part of the job is recognition and refinement. The rest is use. (St. Louis Model City Agency Status Report 1, December, 1966).

Lipsky's (1967:22–28) frame of reference can be meaningfully applied to interpret the Model City Agency phenomenon.[14] The Model City Agency Steering Committee represented various target groups. The proposed program of the Model City Agency would serve to protect target groups from opposition by low income pressure groups. This was designed to prevent riots—to depoliticize revolutionary movements in the Negro community. Protest and violence against business or government was to be directed into a social services model, and playing games. In addition to the downtown HDC bureaucracy for managing tensions between target groups and neighborhood protest, a new bureaucracy of "kept" leadership would be available to "coach" the low income groups. Expertise to conduct demonstrations against exploitation is therefore subverted. Several tactics which protect target groups can be mentioned here. Lipsky (1967: 22) refers to the dispensing of symbolic satisfaction: "Appearance of activity and commitment to problems substitute for, or supplement, resource allocation and policy innovations which would constitute tangible responses to protest activity." The Game of Trade-Off exemplifies this tactic. Besides symbolic gratifications are material satisfactions which may take the form of subsidized employment, food, clothing and the like. Another tactic is to proclaim lack of funds for improving the lot of low income people.

This is a familiar theme of the city administration. Because the Model City Agency, as of October, 1967, had not been funded by HUD, this was an excuse for inaction. However, the hope of things to come is continually bannered by the daily news while action is postponed. The myth that something will be done, slums will be eliminated, and the Negroes deghettoized has become banalized.

JOBS

Jobs eventually came to the forefront as the basic social welfare. Again the Mayor's Office assumed significant proportions as a brokerage but was limited in key aspects such as fund raising and, of course, in providing the jobs. Those receiving the major acclaim were none other than Civic Progress influentials. In spite of the thrust to find jobs and to place people provided by the Human Development Corporation, the Department of Labor, Work Opportunities Unlimited, and many other agencies, the results were dismayingly poor.

In the summer of 1967, St. Louis received five million dollars for a North Side Concentrated Employment Program. By the first of 1968, only 686 had been placed in jobs ("Bias in Job Finding Programs Here Denied," *St. Louis-Globe Democrat,* January 15, 1968). In March, based on 16,000 applications, only 1,100 had received work and about 170 were in on-the-job training (Woo, 1968a).

A Labor Department study made in 1966 found the North Side area, a black ghetto of about 171,000 persons, had 38.9 percent—more than 66,450 persons—subemployed (Woo, 1968a).

Alfonso J. Cervantes (1967:55–65), in an article written for the *Harvard Business Review,* said it was up to business to move if St. Louis was to keep its riot free record. He noted although formerly holding Puritan tenets, since becoming Mayor his outlook changed:

> Observing the riots of Watts (and now Newark, Detroit, and other Harlems throughout the country) has converted me to an updated social orthodoxy. As a public administrator I have discovered that the economic credos of a few years ago no longer suffice; I now believe the profit motive is compatible with social rehabilitation . . . that it is primarily up to private industry, and not to the government, to up-grade the disadvantaged, to provide training for the unemployed, to break down the complexities of job components, to employ the willing, to make them able, to push for social betterment . . . all within the framework of private enterprise, and meaningful democratic government.

He saw no alternative to bona fide business action but arson, riots and a super Welfare State. If riots are to be prevented, then business must bridge the gap, Cervantes commented.

Some recapitulation is needed to understand the job predicament in St. Louis. Work Opportunities Unlimited, the Civic Progress spin-off, by September of 1966 had received more than $300,000 in federal funds. Civic Progress set it up with a small investment, the required 10 percent matching, receiving OEO funding through HDC. WOU's purpose was to provide a job bank for the poverty and employment agencies, and to supervise on-the-job training. This meant utilizing their influence to recruit jobs. Entry level jobs were scarce; many jobs listed were above available skill levels. Also, despite WOU's influence, businesses did not yet have the incentive to modify qualifications. McDonnell Douglas rejected three out of four applicants.

WOU officials accused HDC of poor management of the manpower programs. The confusion about job orders from the Chrysler Assembly Plant is illustrative:

> As WOU originally reported to Civic Progress, it had developed 300 job openings, all paying $3 an hour or more, at the plant. The jobs were cancelled and removed from the Job Bank, WOU reported, because the HDC said the plant was too far away for Negroes living in the ghetto. The cancellation was mentioned in a report on WOU activities that was circulated within Civic Progress. This report played a large part in the businessmen's decision to set up their own employment program.
>
> WOU subsequently submitted another report, which said the order for 300 jobs at the Chrysler plant was placed June 8, 1967, and was reduced Nov. 22 to 10 jobs. WOU reported that in that period 82 persons were referred to Chrysler. Nine persons were hired. In the same period, the report noted, Chrysler had interviewed 11,000 persons and had hired 600.
>
> HDC records, however, 106 persons were referred against the order and eight were hired. In September, WOU received an order of 25 additional jobs from Chrysler, HDC records show, and 21 persons were referred to these jobs. One was hired. On Nov. 22 a third order, this one for 10 jobs, was placed and 25 persons were referred. None was hired. This made a total of 152 referrals against orders for 335 jobs at Chrysler.
>
> "By the ninth of November, our staff already had expressed doubt that the first order was viable," Gatlin said (he is director of HDC work programs). "We recommend it be withdrawn because we felt the job order was no damn good. We were frustrating countless people" (Woo, 1968a).

Later on in this same article Gatlin is quoted as saying: "The total input of business and industry to help redeem the unemployable has been extremely small—but not to be ignored. The level of commitment of business has not been deep enough" (Woo, 1968a).

Another conflict was between the Missouri State Employment Service

and HDC, a situation existing since the latter's inception. The most vitu-
perous hassle came with the funding of the Concentrated Employment
Program through the Department of Labor with supervision by HDC.
Job placement responsibility was to reside with the employment service;
this assumed a kind of cooperation which never fully materialized. The
struggle concerned who would really run the manpower programs.

> In many respects the acrimony was almost inevitable. The HDC,
> a new manpower agency funded by OEO, had been given a super-
> visory position over the established local job placement agency, the
> state service, which is an arm of the Labor Department. The details
> of the CEP were agreed on just before the program was to go into
> effect last summer and neither agency had time to organize or train
> adequately its staff, which needed special skills and sensitivities. . . .
> The HDC maintained that the state employment service was
> damaging the program by hiring women for coaches to work in a
> program that had a special emphasis for getting men into jobs. The
> employment service, it was said, had staffed the out-reach stations with
> rural whites who had no experience or sympathy in dealing with
> Negroes from the ghetto. The employment service was deliberately
> slow in providing HDC with reports and records of its work in CEP.
> For its part, the state service responded by saying that women
> coaches were nothing more than a red herring issue, that HDC had
> not complained about them when they were hired. The rural whites in
> the out-reach stations, it was said were merely temporary workers from
> other state employment service offices who had come to St. Louis to
> help until CEP could develop a local staff.
> Occasionally, the rivalry reached comic opera proportions . . .
> Once, when there was a dispute over the right to use the Job Bank,
> a central repository at the HDC containing a list of all job vacancies,
> Charles DeLargy, area manager for the employment service, and sev-
> eral of his staff submitted fictitious names of applicants to HDC, thus
> taking about 30 jobs off the Job Bank's list. The raid on the bank,
> DeLargy said later, turned out to be unproductive. Most of the 30
> jobs, which were to go to CEP applicants had been filled, he said
> (Woo, 1968a).

The director of CEP said that so much bickering caused them to lose
sight of the program's objectives.

Tensions such as those described in addition to accusations of not
being concerned about job discrimination by one of HDC's neighborhood
stations apparently contributed to the subsequent withdrawal of WOU
from HDC.

Activities at the national level were pointing to more massive involvement in problems of hard-core unemployment. Vice-President Hum-
phrey had wired mayors of major cities to set up coordination of youth

employment programs for summer 1968. Mayor Cervantes asked his brother, who was back again teaching sociology at St. Louis University, to become the part-time coordinator for these programs ("Fr. Cervantes Gets Another Call to City's Poverty War," *St. Louis Globe-Democrat,* January 8, 1968). Previously, Father Cervantes had spent two years as his brother's researcher and analyst. According to Father Cervantes, corporations paid lip service to youth employment needs of ghetto children, and the community lacked funds to implement employment programs.

The National Alliance of Businessmen received its charter from President Johnson in January and a new approach to joblessness took shape.

In an interview (Delugach, 1968) with Frederic Peirce, President of General American Life Insurance Co., concerning the role of business, Peirce said:

> We (businesses) are employers of people. We know how to hire, train and manage their productivity. I think business has a tremendous stake in this. If the community doesn't prosper, neither will we. We have a social as well as business responsibility . . . business can't become an eleemosynary institution, we are in competition.

Peirce spoke of a high priority program involving business and government to secure jobs for 3,000 inner city young people in the summer of 1968 and that a $1,000,000 grant was to be sought. According to Peirce, a United Fund type campaign was planned with top executives asking other top executives to make job commitments. The chief reason for the program was to minimize the likelihood of riots.

Peirce, Harry Wilson of Fleishman-Hillard Inc., a public relations firm, and Fred Sprowles, executive director of the Young Men's Christian Association went to Washington for the million dollars. " 'We started on this with the understanding from Washington that there would be money, and there isn't any,' said Frederic M. Peirce. . . . 'The fellow from the Labor Department said 'Money? Forget it!!' " ("City Won't Get U.S. Money for Slum Jobs Plan," *St. Louis Post-Dispatch,* March 3, 1968). Peirce, apparently, was disappointed and irked.

In connection with the organization of this year's effort to find summer jobs, it was reported that Civic Progress approved alloting $50,000 to the YMCA for administering and coordinating this year's program. Agencies to participate besides Civic Progress and the YMCA were: State Employment Service, Chamber of Commerce, Board of Education, Archdiocesan Board of Education, Human Development Corporation, St. Louis Labor Council of AFL-CIO, Joint Council 13 of the International Brotherhood of Teamsters, St. Louis Building and Construction Trades Council, Metropolitan Youth Commission, Model Cities Agency, Health and Welfare Council, United Fund, Federal Executive Board, Land Clearance and Public Housing, Mayor's Office, and the St. Louis County Supervisor's Office.

Finally, what had been expected happened. Civic Progress withdrew from HDC. The decision was made in February to make a major commitment to the community's National Alliance of Businessmen. Civic Progress, in addition to $50,000 for youth employment, put in $250,000 for the NAB program. The $300,000 came from 24 assessments ranging from $3,000 to $36,000. They pledged themselves to the task of recruiting, training, and hiring Negro unemployed. The heart of the problem is employment and the income which comes with jobs. In order to be effective, Peirce said they would need about $2 million annually in federal funds.

Peirce explained dropping HDC in this fashion:

> This thing got so fouled up by conflicts between agencies which cluttered up the process of getting men to jobs. . . . We thought if we could develop a device that could find people, precondition, train and finally place them on jobs, we could control all the steps in the process. My own description of this is that we could do a quality rather than a quantity job (Woo, 1968b).

At the time of this front page story HDC had not been informed that they had been stricken.

When Peirce became President of Civic Progress in May of 1967, he appointed a civil rights committee including, in addition to himself: Harold E. Thayer, President of Mallinckrodt Chemical; Ashley Gray, President of General Steel Industries; Joseph Griesedieck, President of Falstaff, and Theodore R. Gamble, Chairman of Pet, Inc. The committee met with sociologists, Negroes, education and police officials. From these meetings, Civic Progress became convinced that it could do the job better than the poverty agencies. The newspaper reporter noted that Civic Progress was stepping out of its ordinary role which is to work behind the scenes through other agencies.

The statement of principles in the Civic Progress documents included the following:

> In addition, the experience and background of members of this organization indicate we can be effective in encouraging the formation and growth of more Negro-owned and Negro-operated businesses. . . . There are too few such businesses in the St. Louis area now. Our members have already provided technical assistance and management counsel for several enterprises. We intend to continue this activity. While we have no intention of ignoring other aspects of our racial problem, we believe business should concentrate on improving the position of the employable person, whether presently unemployed and untrained or employed but in a position less than his capabilities jus-

tify, and on encouraging Negro businesses to form and grow (Woo, 1968b).

The plan for a new agency was drawn up by W. R. Persons, Chairman of Civic Progress and was to be submitted to Leo C. Beebe, Ford Motor Vice-President and Executive Vice-Chairman of the NAB ("St. Louis Business to Wage Own Battles in Poverty War," *St. Louis Globe-Democrat*, March 7, 1968). (Henry Ford II, is National Chairman.)

In a *Post-Dispatch* editorial ("A Local Effort on Summer Jobs," March 8, 1968) it was suggested a broader based appeal might be well received in the community, and urged that Mr. Peirce become a repository for contributions. The federal government is chastized (alluding to the one million that St. Louis did not get) for not carrying out its end of the bargain.

This alliance with the federal government is indicated in terms of the announcement of the Missouri Department of Employment Security assigning a recruiting officer to work with NAB in St. Louis. Civic Progress expected to receive money from JOBS, the federal program which is associated with the NAB effort. This recruiting officer was until recently the plant manager of the American Car and Foundry Co., which is a subsidiary of a Civic Progress Co. Under JOBS, businessmen can be paid up to $3,500 annually to hire each hard-core unemployed person. One of the problems before had been employers often wished to avoid the expense of training those deficient in skills ("Urges Re-examining Area Job Programs," *St. Louis Post-Dispatch*, March 8, 1968).

The general manager of HDC praised the new business program to hire the unemployed and said that his agency would give the former its utmost support.

Peirce said:

> Business has the "bread," it has the jobs, it has the personnel, it has the moral and pragmatic incentives. . . . The time for experimentation is past. . . . Providing jobs for the hard-core unemployed will not of itself end poverty in any area. . . . No one can state unequivocally that this vitally important experiment will work, that jobs can be found for, and held by, the currently unemployable. But everyone knows that the attempt must be made, and made with great energy. ("Job Drive Started in City," *St. Louis Post-Dispatch*, March 22, 1968).

The following month, it was announced that a new agency to find jobs had been formed. The group was called BYU representing a coalition of Business, Young Men's Christian Association, and the Urban League. The organization replaced WOU. W. R. Persons, chairman of the board of Civic Progress, chief executive officer of Emerson Electric and past president of WOU said ("Group Formed to Find Jobs," *St. Louis Post-Dispatch,* April 11, 1968):

> Our objective has been to put together a new fully streamlined, strongly managed operation. We hope that BYU, although small, will serve as a model for St. Louis and, perhaps, for the nation. We have arranged for initial funding, and approaches have been made to obtain financing over a longer term. We are already pulling together the necessary personnel and facilities. We are not setting up to take over existing job-finding and job-placing activities. . . . We intend to co-operate fully with other groups, both voluntary and public, under the general sponsorship of the National Alliance of Businessmen.

Later in the month it was reported 629 jobs had been pledged by companies in St. Louis. James S. McDonnell, Chairman of Region VII of the National Alliance of Businessmen, expressed pleasure with the progress. In addition to McDonnell, the following were listed with their positions in NAB: Frederic M. Peirce, St. Louis Metropolitan Chairman who has two Vice Chairmen: Richard A. Goodson, President of Southwestern Bell, St. Louis area, and W. Ashley Gray, Jr., President of General Steel Industries, the Illinois area. In addition, Harold E. Thayer, President and Chairman of the Board of Mallinckrodt Chemical, was Chairman of the job pledge campaign and Monte E. Shomaker, Chairman and chief executive officer of Brown Shoe was Vice-Chairman. Loaned full-time was William M. Holland, Executive Supervisor, Southwestern Bell ("629 Jobs Pledged by Industries Here," *St. Louis Globe-Democrat,* April 22, 1968). It was the UF all over again.

In an editorial ("The NAB Means Business," *St. Louis Globe-Democrat,* April 27–28, 1968) the *Globe* said that not since the depression has there been such a challenge.

> Business leaders across the nation are in the forefront of the fight for jobs—with good reason. They are responding to a plea from President Johnson which they realize they could not reject. The President's request was in itself a tacit admission that business, rather than government, has the brainpower and capacity to meet the task. Politicians have failed to solve the problem. Program after program under federal auspices has resulted in overlapping agencies becoming top-heavy with overpaid administrators. Meanwhile the poor remain poor and out of work. It has been said that business leaders represent the nation's "last hope" in the current crisis. We believe this to be true. It is no exaggeration to say that the American free enterprise system has its finest opportunity to prove its worth to the entire world.

The foregoing information about the drive on jobs needs little interpretation to assert the influence of the business community both locally and nationally, and the denigration of the polity.

The Mayor's Office served as a broker, but there was no question about where the real power resides. In the subsequent campaign to raise money for various kinds of summer programs, the Mayor's Office had the slick brochure called "Push Three, Employment, Education, and Recreation," but the envelope for mailing the contribution had the following address:

> Push 3 YOUTH—SUMMER '68
> Mayor's Council on Youth Opportunity
> c/o CITIZENS' ALLOCATION COMMITTEE
> Health and Welfare Council
> 417 North 10th St.
> St. Louis, Missouri

Again, the Mayor did not collect the money.

CONCLUSIONS

De jure private agencies, such as the East-West Gateway Coordinating Council and the Human Development Corporation, which are de facto dispensers of public funds, constitute models for destruction of public government. Instead of such structures being responsible to the electorate, they are tongs of urban power by which civic influentials can manipulate social welfare. Although appearing as servants to federal and nonfederal polities, EWGCC and HDC are not in fact, because they depend upon interlocking organizational interests of regionally based economic wealth.

Efforts to break the private ideological pattern of downward mobilization of action were continually thwarted. Upward pressure for change was absorbed by statutory voluntary agencies such as EWGCC and HDC which have their being in statutes, but are voluntarily instituted and significantly controlled by interests not responsible to an electorate. In the case of HDC, it was rank preemption of protest leadership. The placement of the Metropolitan Youth Commission under the wing of the Health and Welfare Council positions the former so that it is more controlled by civic elites. The resistance of Civic Progress influentials to greater democratization of hospital planning was manifested in their having to be coerced into following certain recommendations of the USPHS.

The campaign for Medicaid indicates close cooperation of private and public influentials at local levels in the state to pressure the governor. This was done to save cities money, and as pointed out, General American Life Insurance Co. made a business deal out of the transaction. It should also be noted that health care was extended.

The Model City Agency, although structurally under the Mayor's

Office and therefore a part of city government, is de facto private in the sense that the expertise has come largely from business and private community agencies such as the Health and Welfare Council and the Metropolitan Youth Commission. It is ironic that whether an agency is de facto private or public, its subordinance to the private sector is assured. As with HDC, the Model City Agency functions to protect target groups such as business and legitimating institutions from attack by blacks.

The National Alliance of Businessmen represents a superlative evolution, thus far, of the structural model. Here the agency is dominated by businessmen in a de jure as well as a de facto sense. This is tantamount to self-made law—only instead of an individual, it is a collective doing this. Being chartered by the President, NAB has the highest imprimatur that can be afforded. National and local business influentials command on their own terms. This gives them the right to use the Post Office franking service. But this is perhaps not too startling when one is being prepared already for the take-over of this Department by a private corporation.

The take-over of guidance, training, and educational functions portends further decline of polity. People have lost confidence in their abilities to govern. Thus, it is only natural that more sophisticated organizational models would outdistance public governments. The private ideological model in St. Louis is authoritarian and totalitarian, instituted with public acceptance, and ensconced in the rhetoric of law and order.

The effectiveness of this model is attested to by the fact that, up to this point in May, 1969, St. Louis has not had a major riot. As a pacification system, Civic Progress in large measure makes it work.

CONCLUSION

The central problem is who controls the organization and ideology of voluntary agencies, and has this control been lessened by the rise of public ideology? The purpose of a case study is to elaborate and develop hypotheses for further testing. The tentative answers given are: (a) Influential private agencies control interagency organization of social policy, and (b) Influential private interagency control of social policy is scarcely lessened by the rise of public ideology. Influence was studied in terms of relative power inferred from the superordinance and subordinance of external lines of power and interactions of organizations. The above-mentioned hypotheses have been progressively refined during the course of the research. Answers to questions and major lines of evidence concerning control and ideology are presented in this conclusion.

(1) Large locally based corporations have interlocking boards of directors.

This provides, as Mills noted, a community of interests and facilitates consensus in business decisions. Because such companies are interlinked, opportunities for reciprocities are enhanced. They buy from each other in a process of mutual enrichment and they protect their economic territories from intruders. Lesser companies depend upon the largest for sales of products and services. The interlocking of corporate roles is congruent with Hunter's (1953) and Mills' (1956) theories on dominance of the economic institution and corporate elites in the community's power structure. They are finpolities.

(2) Corporations have overlapping memberships on the boards of community agencies.

Boards of voluntary agencies, like their corporate counterparts, have interlocking directorates. This fosters consensus, also, in decision making and contributes to the domination of social policy by business influentials. Because boards of community agencies are self-perpetuating, executives from these corporations can nominate each other as they rotate off one board of directors on to the board of another agency. Many corporate

181

executives hold multiple board memberships; some boards have no policies concerning how long a member can serve.

The foregoing is consistent with Hunter's findings concerning the interlocking of club memberships in Atlanta.

(3) Chief executive officers of corporations join to form a supraorganization for control of social policy.

In St. Louis, this supraorganization is Civic Progress, Inc. This association facilitates a more efficient mechanism to formulate and execute community policy than merely interlocked directorates. Major functions of this executive committee of the corporate elite have already been mentioned. They may be briefly summarized in the following:

Civic Progress is the supreme power in the community for purposes of philanthropy and fund raising, be it private or public. This comes from having a monopoly of corporate wealth and the means for raising money in the community. The preemption of the United Fund by Civic Progress blatantly exemplifies totalitarian control of community services. By its domination, it determines how much money will be raised, allocation of funds, which agencies can be admitted and/or excluded as unworthy of the community's trust. To minimize duplication of effort, Civic Progress regulates campaigns for solicitation of funds. By banding the largest corporations together, payroll deductions are restricted to the United Fund which sets up constraints on private philanthropy not included in the United Fund. In so doing employee giving is dictated. Besides raising money for charitable purposes, they were money getters for the Civic Center, Spanish Pavilion, Summer Youth Programs and others. The Mayor's Office depends upon these civic stalwarts for support in getting public improvement bonds passed, and Civic Progress was instrumental in the passage of increases in the earnings tax for the city.

Another attribute of Civic Progress' influence is that it creates other agencies. Being an agency maker is usually contingent on access to fiscal resources. Large corporations and the federal government have these resources or possess the power to obtain them. The Mayor's Office in the City of St. Louis lacks the financial reserves to establish agencies, and has less freedom to do so because of a closer relationship to the electorate than either business corporations or departments of the federal government which also set up private agencies to carry out governmental programs. Civic Progress, for example, created the United Fund, Hospital Planning Commission, Downtown, Inc., and the Regional Industrial Development Corporation. The new agencies such as the Human Development Corporation and the Model City Agency, which are private agencies spawned by the Office of Economic Opportunity and the Housing and Urban Development Department respectively, needed and received endorsement and consultation from Civic Progress members, as well as agencies under the domain of Civic Progress.

Civic Progress has the unparalleled capacity to study community problems and to recommend action. In many matters Civic Progress and its members are beholden to no others but themselves, and thus their actions can be swift and sure. Because of money and manpower at their disposal, expert consultation can be obtained. Where not available in the corporations, the expertise of the policy sciences can be bought from either the local community or elsewhere. Control over facts adds to Civic Progress' power to influence strategic decisions. Because of staff of lesser talents and limited budgets for consultation, city departments and agencies depend in part upon corporation personnel or consultants employed by Civic Progress.

Men of Civic Progress are promoters par excellence. This is evidenced by their campaigns to change the city charter, promotion of bond issues, tax increases, urban redevelopment, civic attractions, the arts, and the private and public educational systems of the community. Recently, they were active in persuading Governor Hearnes to implement Medicaid. Organizations under their control, such as the Health and Welfare Council and Chamber of Commerce, lobby for legislation or can be discouraged from supporting bills not agreeable to Civic Progress.

In addition to bolstering the area's economy, Civic Progress members are elected to important posts of civic responsibility, have ready access to the media for their pronouncements, and are continually honored by different groups in the community. Hence, Civic Progress has the imprimatur of top leadership which projects an image of god-like stature in the upper economic circles of the United States and throughout the world. Mills, Hunter, and Presthus described the national leadership qualities of business influentials. St. Louis is a national city and in this regard is not unlike other large cities. For August A. Busch, Jr. and James S. McDonnell to be personal friends of President Johnson is a critical fact in understanding a community's power structure.

(4) The sectarian organization of social welfare is integrated through the corporate power structure.

Civic Progress represents the Protestant Establishment even though it incorporates Jews and Catholics in high level decision making. It is perhaps because of the lack of a viable ecumenism that sectarian voluntary social welfare can be managed in a secular process. Organizational anarchy, both within as well as between sectarian associations, creates a power vacuum filled by economic influentials. Pluralism, in this sense, fails to safeguard sectarian autonomy but seems to contribute to domination by corporate welfarism. Baltzell's (1966) hypothesis that social class substitutes for religion and ethnicity in the upper stratum is largely confirmed. However, the day has not been reached—in St. Louis at least—when Protestant industrial power has been distributed irrespective of creed or ethnic group.

The facts about Civic Progress as a constellation of power elites, do not fit the pluralist notions of Dahl (1961), Long (1962), Banfield (1965), Rose (1967), Greer (1963), Bollens (1961, 1965), Schmandt (1961) and their adherents. Chief executive officers should, according to the pluralists, confine themselves to social welfare as a substitute for political controversy with which they were associated during an earlier period when business and politics were closely allied. Moreover, they should be prominent only in one major field, according to Wildavsky (1964). In St. Louis, the pluralist hypothesis does not coincide with the evidence on community lines of influence. These men, as well as their supraorganization which interpenetrates other organizations and creates specialized organizational roles in spinning-off new agencies, operate in diverse fields of politics, social welfare, the arts, education, and civic development.

(5) Corporation centered welfare is idealized as having top priority for life contingencies and needs.

Earnings are given primary emphasis in securing the individual's well-being. This is the Protestant Ethic which underlines frugality, hard work, and personal responsibility. Pension plans, health care, profit sharing, deferred income payments are indicative of this kind of welfare.

(6) Private community welfare is a second line of defense.

United Fund giving, Old Newsboys, Neediest Cases, Arts and Education Council campaign, gifts to St. Louis University and Washington University, and many other philanthropies represent the most common subtype of private community welfare. Limited profit ventures such as the Civic Center and Spanish Pavilion fit this brand of welfare. The activities in Work Opportunities Unlimited and later in Business, YMCA, and Urban League (BYU) in connection with the National Alliance of Businessmen are another subtype.

(7) Public community welfare is idealized as residual by corporate influentials.

The emphasis is on a modicum from the minimum wage, public assistance legislation, unemployment compensation, social security, and other types of governmental programs. This is congruent with the Protestant Ethic, and with the conviction that the source of well-being should be from corporate welfare and not from the welfare state. It is, therefore, apparent that the established public welfare programs administered by statutory agencies have been penetrated by private ideology. In spite of the publicity received by the Human Development Corporation, virtually all welfare money is spent through these kinds of agencies whose wings are clipped by statutes written in the true spirit and letter of the Protestant Ethic.

(8) The growth of a disadvantaged Negro population (with the concomitant Negro protest) stimulates public ideology.

During the 1950s and into the early 1960s the Negro population in the city was rapidly increasing and spreading. Whites were losing political power to the Negroes as the former's standard of living escalated with the rising fortunes of the corporations, enabling them to move into the virtually all white St. Louis County. Civic Progress was initiated to cope with the declining tax duplicate, downtown decay, a city which had become substantially dilapidated, and whose school children were mostly black.

Federal programs and funds were available through public housing and urban renewal and Congress attempted to legislate Negro freedom and equality through fair housing, fair employment, and public accommodations. Public ideology gained a feeble foothold. However, public as well as private improvements did not seem to benefit the Negroes where they needed the most help: Negro family incomes remained substantially lower than family incomes of whites. Help for the Negro bogged down in the public and private agencies pervaded with private ideology and did not secure social justice because it failed to provide him with adequate income which is essential for the Negro to raise his social status.

(9) Private ideology fails to extend the benefits
    of the corporate system and creates private
    and public welfare which only recently have been
    directed toward increasing Negro incomes
    to any appreciable extent.

Hence, private and public welfare programs were hoisted on their own petards. The private welfare system was accused of being ingrown and serving best the accessible, the willing, and commonly the middle class. This is documented by S. M. Miller (1964), Sanford Solendar (1963), and others.

(10) Statutory voluntary agencies depend
     upon the locally based corporations
     for implementation of programs.

These agencies are de jure voluntary but de facto statutory since programs to a great extent are stipulated by federal statutes. They are instituted on a voluntary basis to be outside the existing organization of social welfare—autonomous from private and public agencies. Their independent status, it was hoped, would enable them to remove the institutional barriers to Negro self-realization.

The Human Development Corporation is such an agency and serves as the federal government's "trojan horse." Although originally attacking the institutional structure of both private and public welfare, it soon incorporated the leadership of the nonstatutory voluntary agencies and the public agencies in the War on Poverty. The nonstatutory voluntary agencies have a monopoly of social services and trained social workers needed by the antipoverty programs. Moreover, the employment programs required the cooperation of the corporations. Giving token investments to job programs initially, they apparently performed supremely on behalf of their own organization, the National Alliance of Businessmen, chartered by the President of the United States. In the process of shifting from WOU (Work Opportunities Unlimited) to BYU (Business, YMCA, and the Urban League), Civic Progress manifested its contempt for the Human Development Corporation by letting the latter find out through the newspapers that it had been dropped. Of course, all along, participation by the voluntary sector in HDC programs had been one of studied restraint.

The community action program to change the bureaucracies of the public and private agencies was crushed in the early stages by city and state public agencies which operate in the local community. This happened in other cities as well as St. Louis. The political action program of the civil rights movement became oriented to social services. Education, housing, legal, and employment services became substitutes for action which could lead to income sufficiency. Civil rights leaders were employed by the new bureaucracy of the Human Development Corporation which had found its place in the hierarchy of community agencies. HDC had embarked upon "the numbers game" like the Missouri State Employment Service. The goal was job placement whether poor paying or not; whether make-work or not or regardless of the duration.

The preceding account of events supports the theses of Rainwater and Yancey (1967), Lipsky (1967), Horowitz and Leibowitz (1968) that the political thrust of Negroes is converted into a social welfare model. Service strategies rather than income strategies became the goal and the method.

(11) Public ideology is subverted by the voluntary and
      public social welfare system. Statutory voluntary agencies
      submit to the nonstatutory and public agencies.
      Public ideology, such as it was in the
      Economic Opportunity Act, when operationalized,
      is dominated by private ideology. The myth of
      public ideology remains, but the reality
      is a private ideology serving the corporate
      power structure.

Because problems of low income are unresolved, riots provide the stimulus for new programs and new panaceas. Also the Negro vote of the large cities is at stake. The Model City Agency under the aegis of the Mayor's Office is to be founded under the Demonstration Cities Act. The voluntary and public welfare establishment have served well as midwives for what is promised to be far superior programming than what preceded. Income was to be the central criterion for program success, but in view of recent proposals regarding housing, neighborhood centers and transportation, income as a strategy seems safely ensconced in innocuous service programs which promise higher income with no direct threat to the corporate profit system.

It can be contended as a counter thesis to the one presented that: (a) Business influentials are naive about social welfare; that they are powerless in coping with the major problems of urban life because local problems are national problems; (b) Since they are powerless, they are forced not to develop rational understanding of these problems; (c) They maintain their image of being powerful by engaging in symbolic behavior vis-à-vis social policy—a rational stance, in lieu of the myth of problem solving, would reveal their impotency in dealing with crucial issues such as full employment and adequate income maintenance; (d)

They have two major interests—achieving profits and social prestige, and should not be dealt with too harshly for civic welfare activities which may consume less than 10 percent of their time. They need to spend a minor amount of time in validating the imputation of power; (e) They really do not care about the plight of the poor—except that because of their community statuses the elites are forced to care; (f) The urban riots are not perceived by the business elites as a serious threat to their holdings; (g) They do not oppose liberal reforms—rather they are either naive about the causes of social deviance, or social welfare policy making is marginal in importance to the corporate power elites; (h) Thus, they are merely figures of power in a community's folklore—demoniacal to those who oppose the large corporations, and simply needed and used by politicians, administrators of educational institutions, social workers, and sociologists to play games which mobilize various publics to carry on the civic life. There is illusion of power but not the real power. These are likened to supernatural creations of primitive communities—a part of the civic magic. There is a need to believe in city fathers. According to Long (1958) the ills of society are inconsequential so that the games do not matter a great deal: "Much of civic ills also cure themselves if only people can be kept from tearing each other apart in the stress of their anxieties. The locusts and drought will pass. They almost always have."

How naive corporate influentials are about community welfare problems is a matter for precise conceptualization and empirical research. If Civic Progress may be taken as an indicator of business elite's intelligence quotient, one would be inclined to surmise they are among the most informed about the broad sweep of community problems. The fact that they have not removed themselves from civic concern means they continue to be in a position to have the benefit of information directly from the technocrats of the community and consultants near and far. Moreover, to delegate responsibility to professionals and to other subordinates, as they do in the organizations which they have spun-off, does not necessarily result in naïveté concerning those matters which have been entrusted to others. To the social scientist it may seem as though the business elites are naive, but are they actually naive, or instead capable of being informed by experts at points along a critical path of social policy? A top corporation executive does not become naive by depending upon excellent staff work. It would be the acme of naiveté to do otherwise.

Is corporate power actually localized? Is it true that any local elite operates in a system in which it is not preeminent? Are the forces beyond the local communities independent of the influence of community power structures? Hunter's (1959) and Warner's (1967) researches suggest a national business community and power structure. Although more informally organized than the government structures at state and federal levels, the national power structure is generated by business reciprocities. There are also interlocking directorates which are some of the more formal social cement which bond the large private states together. The dependence of government on the nongovernmental organization is noted by Pifer. The technical skills and the means of production are owned or readily purchasable by the large corporations.

There is a difference between corporate elites not having a rational understanding of the problems and having a rational understanding, but failing to act upon what is known. It is possible that their approaches are rational in terms of protecting their holdings. Rationality is of course a question of ideology as well as science.

It may be true that community social policy, particularly as it relates to the disadvantaged, is based largely on myth and playing games. However, to interpret this as impotency of business elites may be due to not taking the role of the latter in the ecology of community games. Such symbolic behavior may be indeed functional for maintaining the dominant position of influential agencies. To democratize the social system; to remove the cumulative inequalities in resources of influence (see Dahl's questions in chapter 1) which come from interlocking directorates and overlapping of agency board memberships might overthrow the monopoly of board positions by business influentials. This monopoly is, of course, in terms of nominations, incumbency, and the absence of popular elections. Moreover, how rational would it be to turn over an empire when it is solidly backed by the public?

The thesis is consistent with the counter thesis in that corporate elites have three major interests: profits, prestige, and stability. Disagreement comes in thinking that social policy making can be separated from business making. The two are interrelated and interdependent. Hence, in the generic sense it is all welfare. Slums are profitable. Urban renewal is profitable. Defense industries are profitable. Medicare and Medicaid are profitable. To keep the blacks quiet makes for stability. To avoid excessive outlays for air pollution control is profitable. For government to pay for the processing of the unschooled and untrained so they can be employed by the corporations is profitable. It is profitable in terms of payoffs to stockholders and in higher salaries to corporate chiefs and their various subjects. The finpolities or private states, besides the time consuming business of their internal welfare programs which are nonetheless social policy, also influence social policy governing the relations among themselves and the remainder of the community domain which includes its vertical extensions in the nation.

It does seem that the corporate elites are forced to care about community problems. Revolutionary elements are in the system which must be dealt with and black power is one of these. If business elites did not show some response to these community pressures, would they continue to be the leaders? Moreover, if they would respond in more socially significant terms as far as really doing something about inadequate income and unemployment, they might not long be the rulers either. Basic conservatism in the public girds the loins of the business aristocrats. Because they do act, they are spared the criticism which might otherwise be inordinate from liberal elements which follow the philosophy that a little bit of social gain is better than none at all. Of course, most intellectuals do not reject the profit system on which the private business corporations rest. Thus, criticism by intellectuals are reasonable and not too unacceptable, as a rule, to the elites who realize that drastic revolutions can be aborted by accommodations.

The black revolution with its concomitant violence shakes the foundations of the republic. It is no doubt the case that the black masses are apathetic, but it also seems correct that these masses did not appear to be active in restraining the rioters—in fact, they often became participants. Therefore, it is important for the economic elites to preempt black leadership; to employ their fire eaters in more creative endeavors. This has been done and will likely continue to be a pattern of influence.

The strong support of business elites for the national Republican Party is indicative of their favoring a policy of status quo. In spite of social reforms and innovative programs, to which American business has been lukewarm, urban disorders are mute evidence of the inadequacy of social policies. Therefore, even though Republican candidates for the presidency and Congress have frequently lost, in the past thirty-five years, many Democrats are hardly more liberal than their defeated opponents. The corporate sector is in the fortunate position of enjoying whatever benefits come from the "socialistic" programs, and eluding the public criticism of the liberal's failure.

No doubt there is a civic magic—an aura or charisma surrounds the business leaders of the community built upon their public benefactions. In this game the public servants are the private servants and the private leaders are the public men. Behind the symbolism is the obdurate fact that people earn their daily bread from these corporate giants. People depend upon the corporate welfare in the primary sense as interpreted by the corporations and as seen by people themselves—that is, for personal income. Moreover, social policy also benefits the corporate structure in the form of paid vacations, health insurance, pension plans, and profit sharing.

Coming back to Dahl's (1961:7) questions again: "What kinds of people have the greatest influence on decisions? Are different kinds of decisions all made by the same people? From what strata of the community are the most influential people and leaders drawn? Do leaders tend to

Appendix A

List of Abbreviations

Used in This Book

ACTION	Action Committee to Improve Opportunities for Negroes
AC	Arts and Education Council
ADC	Aid to Dependent Children
AFL-CIO	American Federation of Labor and Congress of Industrial Organizations
BCIC	Bicentennial Civic Improvement Corporation
BYU	Business, Young Men's Christian Association, and Urban League
Cath Ch	Catholic Charities
CC	Chamber of Commerce
CCRC	Civic Center Redevelopment Corporation
CEO	Chief Executive Officer
CEP	Concentrated Employment Program
CORE	Committee on Racial Equality
EWGCC	East-West Gateway Coordinating Council
HDC	Human Development Corporation
HEW	Department of Health, Education and Welfare
HPC	Hospital Planning Commission
HUD	Department of Housing and Urban Development
HWC	Health and Welfare Council of Metropolitan St. Louis
JCCA	Jewish Community Centers Association
JEVS	Jewish Employment and Vocational Service
JF	Jewish Federation

JOBS	Job Opportunities in the Business Sector
KSG	Knight of St. Gregory
LBJ	Lyndon Baines Johnson
MCF	Metropolitan Church Federation
MYC	Metropolitan Youth Commission
NAB	National Alliance of Businessmen
NAACP	National Association for Advancement of Colored People
OEO	Office of Economic Opportunity
RIDC	Regional Industrial Development Corporation
SCLC	Southern Christian Leadership Conference
UF	United Fund
USPHS	United States Public Health Service
WOU	Work Opportunities Unlimited
YMCA	Young Men's Christian Association
YWCA	Young Women's Christian Association

Appendix B

Highlights of Main Interviews

Both direct quotations and paraphrasing of interview responses are given. The question is stated and then the answer is included in quotes only if it is a direct quotation. Notes were taken during the course of each interview. They were written up afterwards. An effort was made to record what was said in the respondent's own words. The paraphrasing includes a summary of the respondent's remarks; explanations are enclosed in parentheses.

The interviews were informal with certain key questions included. These were varied depending upon the person being interviewed. These were in part focused interviews since certain questions were asked to clarify areas where my experience, or archives, newspapers, and local reports were insufficient for purposes of establishing the facts or making proper inferences.

The interviews are supplementary to the main line of data collection in confirming or disconfirming theories about agency relationships.

As case studies directed toward hypothesis formation and elaboration, this relatively unstructured, but focused, form of interviewing seems productive of spontaneous and insightful responses.

INTERVIEW 1

ROBERT K. SANFORD (*Post-Dispatch* staff member and author of the series of articles on Civic Progress, Inc.: interviewed January 5, 1967, concerning additional information on Civic Progress).

197

Question: What does Civic Progress think about using public money?

Answer: "It depends on the project . . . they are interested in things they can control. They have government contracts, and are used to doing business with the government. . . . They would not want to be known as against federal money. They really consider each thing separately."

Question: What is their major concern now?

Answer: "What has been done by Civic Progress is essentially brick and mortar. Much of this is completed or is on the way. The major issue now is social. Civil rights is seen as the number one problem. They see social services being taken care of by the government, and their contribution can be in the area of employment. They are bending some from the straight business viewpoint by trying to give some of those who are hired a chance to grow up on the job. If they don't make it on one job, perhaps they can on another."

Question: Why did Civic Progress decide to stay small?

Answer: "They were originally going to be a large organization with a staff. Instead, they decided to stay small and to work through existing organizations or set up new ones. . . . They could not get the executive they wanted. Also, they observed that some big community organizations did much talking but not much acting."

Question: Why was Downtown, Inc. set up?

Answer: "Aloys Kaufmann did not think that they could get the support of the Chamber of Commerce membership which includes many businessmen who have no downtown interests."

Question: What about the Regional Industrial Development Corporation?

Answer: "This came about because of the conservative influence of the Chamber of Commerce. Some retailers could be opposed to bringing competitors into the area."

Question: What is emeritus membership in Civic Progress?

Answer: "This is a nonvoting membership. Some of these guys feel that they have more to offer than the ones who are currently active." (He gave examples of Edwin Clark, previously chief executive officer of Southwestern Bell Telephone Co., and Ethan A. Shepley, previously Chancellor of Washington University and a former gubernatorial candidate.)

Question: How powerful is Civic Progress?

Answer: "It is still a force. There is so much money represented by those corporations. The banking interests in particular are closely tied together here and with Kansas City. There is a close tie in with the Kempers and Cravens." (The Kemper family dominates the banking community in Kansas City, Missouri, and Cravens is the head of Mercantile Trust Co. of St. Louis.)

Question: Why isn't labor in Civic Progress?

Answer: "Harrington (President and Board Chairman of Boatmen's National Bank) told me that they didn't know who could speak for or how much labor's viewpoint was needed."

Question: How powerful is labor?

Answer: "Morton May tried to get the Arts and Education Council in the UF, but some of the business people in the UF thought that labor would never buy it. . . . The success of the UF depends on labor's support."

Question: What about Civic Progress' relation to city and county governing bodies?

Answer: "They throw a party for the Board of Aldermen at the end of their session each year. They do the same thing for the County Council."

Question: Why isn't J. S. McDonnell (chief executive officer of McDonnell-Douglas) in Civic Progress?

Answer: "He was invited, but he sent a vice-president, and they told him to forget it. They have his brother anyway." (William A. McDonnell, chairman of the finance committee of the St. Louis-San Francisco Railroad.)

Question: Any conflict between Civic Progress and the St. Louis Ambassadors? (These were appointed by the Mayor to help promote St. Louis.)

Answer: "Hell no. You don't name a building after the guys who only put on the top floor."

INTERVIEW 2

DAVID R. CALHOUN (Chairman of Civic Progress and President and Board Chairman of the St. Louis Union Trust Co.: interviewed March 9, 1967, concerning what is happening to voluntary agencies).

Question: What is happening to voluntary agencies given the power of public funds?

Answer: "Agencies may be divided into two groups: those which receive public funds and those which do not. Everybody is trying to get these federal grants and the government is stimulating it. This intrusion of the federal government is a political thing. These federal funds keep pouring in—nobody has got the guts to stand up and say no to these things. Conditions have gotten to the point in the city where there just doesn't seem to be any other way. The trouble is that people are in a hurry today—they can't wait to make it. It's part of the socialistic trend in the country. Besides the cost of the thing, it tells people that they'll do things for them that they can never do—such as eliminate poverty. You don't know what this will do to ignorant people—how they will react. There's the Communist influence out in Berkeley, and I just don't know how much that had to do with the Watts thing. The trend is for more and more government intrusion, and I don't think it can be stopped. . . . What was done for the Herbert Hoover Boys' Club is different. Somebody suggested that public funds could be obtained to buy something that was needed. I said: 'God damn, we've done the whole thing with private money up to this point—let's keep it that way.' "

Question: Why have the UF allocations increased only three percent per annum since 1960?

Answer: "It isn't enough is it? Some of the agencies are so expensive to operate. The per capita costs of some day care centers are high. There is also the matter of making the goals. Sometimes we set them too low, I realize. But it is so important to make the goals, and for the image of the community. . . . It is a helluva way to raise money for the agencies and institutions. You can't beat it in terms of the cost of raising the dough. . . . Most give willingly but you have to twist a few arms."

Question: What about using UF money for matching funds for HDC programs, even if matching requirements go up to twenty percent—will we let the federal money go?

Answer: "We won't do that. It is a helluva good deal—it's too good to turn down."

INTERVIEW 3

JOHN MYRICK (former Campaign Manager of the United Fund: interviewed January 6, 1967).

Question: What about all the campaigns in the community?

Answer: "They are of concern to the executive committee and agency relations cabinet. However, federation is a voluntary matter. These drives continue because people continue to work in them and give to them. The Easter Seals is a very expensive campaign in terms of the returns on what is sent out. These health outfits get very little more than token contributions from the big corporations."

Question: What do the business leaders think about the War on Poverty?

Answer: "These are practical men. Even though they are conservative, they will support something when they know there is nothing which they can do about it. This is the way they run their own businesses. If they see that something is losing money for them—or they could make more money by closing out a product line—they will act. The same way with the hospitals. They would prefer to continue their support, but since the government has gotten into this, they are simply forced to get out because it is just good business sense. They figure that some one needs to be in there to see what is going on."

INTERVIEW 4

WILLIAM KAHN (Executive Director of the Jewish Community Centers Association: interviewed January 3, 1967).

Question: What is the future of voluntary agencies?

Answer: "We can expect more governmental funding of all kinds of welfare programs. . . . But the needs are great, and there is still much room for voluntary agencies to be innovators. Also there is a need for an outlet for philanthropy. The psychological reward in private philanthropy is still great. It is not weakening—fund raising is stronger than ever. Multiple fund drives give many people a sense of goodness, humaneness. . . . The Chamber and Fund may be concerned, but on the whole this is a healthy sign of community life. The Jewish community has a hundred drives going on. . . . There is some thinking that when the "fed" money comes in, the private money should be pulled out."

Question: Any difference in political outlook between affluent Jews and gentiles?

Answer: "The affluent Jews are very conservative—no different than the gentile affluent in worrying about the free enterprise system."

Question: What about control that may come with accepting public funds?

Answer: Mr. Kahn indicated that he is not troubled with controls that can come with receiving federal money. Federal funds are needed for special services which could not be financed except through public money. "There is a lot of red tape and control with the private dough also. There have been few strings with the federal money that the Center (JCCA) has had thus far. There are a lot of stereotypes on the business of taking fed money."

Question: Does JCCA have the top leadership of the Jewish community?

Answer: Kahn mentioned JCCA has a Board of Trustees and Irving Edison, in Civic Progress, is on that board. "They are the board behind the board . . . an informal group of key guys who will backstop when necessary."

Question: What about the Jewish Federation?

Answer: "It has not been as successful here as in other communities—not the commitment of the corporate elite among Jewry. . . ."

Question: How powerful is Civic Progress?

Answer: He says that Civic Progress has considerable influence in the community. "When JCCA needed to get contributions from the gentile community, Russell Dearmont got some of the guys together who kicked in what was asked."

INTERVIEW 5

WILLIAM DOUTHITT (Executive Director of the St. Louis Urban League: interviewed January 10, 1967).

(He was in the midst of a capital funds drive to pay for the building into which the Urban League had moved. He said that it was "tough going." He mentioned working with Wesley McAfee, Chief Executive Officer of Union Electric, and G. S. Roudebush, Vice President and General Counsel for McDonnell Douglas.)

Question: What about Civic Progress?

Answer: "These men have power because they control the money in the community . . . They will let you do pretty much what you want in the way of social welfare as long as you don't interfere with their money making."

Question: How is Work Opportunities Unlimited doing?

Answer: Mr. Douthitt described it as a "protective device" and that "management criteria" are being used to run it. It is competing with the Urban League.

Question: How is citizen participation in the antipoverty programs working out?

Answer: "Maximum feasible participation of the poor is a pain in the ass. . . . These people try to take over the program. They want to be given the money to spend for themselves. . . . When I tell them that they do not have the knowledge to run these things, they call me a Caesar."

Question: What about Gibbons' influence? (Harold J. Gibbons is head of the Teamsters Joint Council.)

Answer: "He is a force politically . . . a career diplomat . . . who has made a kind of accommodation with business. . . . They are getting no place trying to organize Famous-Barr and did you see how quickly Egan stepped down from the firefighters when they went on strike? (Egan was the head of the local and also a full-time paid labor representative for cohere in their policies and form a sort of ruling group, or do they tend to divide, conflict, and bargain? Is the pattern of leadership, in short, oligarchical, or pluralistic?" The chief executives of the corporations in Civic Progress, through pivotal community agencies, would seem to have disproportionate influence on decision making. The same chief executives and their corporations are making different kinds of decisions regarding charity, education, the arts, urban renewal, bond issues and the like. Business, civic, and political spheres are kaleidoscopic. Civic Progress meets Dahl's stipulation as a ruling group. Norton E. Long (1962) notwithstanding, it seems to be a veritable executive committee of the bourgeoisie.

Finally, the question of who rules can probably only be fully understood when analyzed in ideological terms. The civic ills are no longer easily palliated and concealed. They do not cure themselves, and to say they do is to reflect an ideology which would make individuals adjust to a structure of power rather than to democratize that structure.

For the pluralist hypothesis to be confirmed, there must be differences in ideology mustered by a plurality of voluntary organizations which mesh debating publics with decision making centers. Instead of polemics about social justice, there is consensus with (more or less) acceptance of minimal standards by those who might be thought of as public ideologists. The wealth of the business corporations, organized labor, the church, the newspapers, the educational system and the general public is committed to the Protestant Ethic and to Corporate Welfarism. Social justice is viewed anemically. Thus public ideology becomes like a feather brushing softly against the grim realities of low income living.

The quality of public ideology needs to be evaluated for its equity content. When this is done, the programs thus far seem to be congruous with the Protestant Ethic and Corporate Welfarism. What was billed as the "Great Society" became the "Little Society" because in spite of being serviced by programs which supposedly manifest a public ideology, many Americans still do not have adequate incomes to meet their families' needs. Thus public ideology is compromised and is public in name only because it imposes an oligarchical rule to exploit the many.

Private and public influence, it is concluded, are used to protect the holdings of the powerful corporations. Social welfare is a substitute for significant changes in the social and economic order. Political deviance is managed by being transformed into the social welfare model. Where deviance is not amenable to social treatment, it is judged to be criminal.

Social welfare also serves to legitimate the business influentials. They are known by their good works which create a favorable image for their profit making that exploits the many. If black protest is warped by the social welfare model, then freedom and equality will be compromised. In an earlier period the farmers and then labor each made their bids for equity and as Mills (1956) points out, each group failed and succeeded. They failed in maintaining themselves as autonomous groups to oppose the corporate rich, but they succeeded, as perhaps the blacks will, in positioning themselves as "middle powers."

The consensual relationships between business and government at city and national levels make for an imposing power conglomerate. What results is totalitarian democracy at the community level realized by the provision of welfare through the polity or finpolity. Political deviance is repressed, and the status quo only appears to change. If this be so, then the cherished multiassociational society of Almond and Verba (1963) is illusory and Rousseau's (1762) general will is the reality.

Voluntary social welfare agencies are beleaguered. Their status is ambiguous, and it has become respectable to ask if they are really necessary. Voluntary agencies have been criticized by social scientists as well as by the stewards of social welfare—the professional social workers. With increasing public subvention of social welfare, and the generation of statutory voluntary agencies, which are in effect "fronts" for polities or finpolities, the existence of voluntary social welfare agencies which do not depend upon public funds hangs in the balance—subject to the noblesse oblige of the corporate rich. The nonstatutory voluntary agencies continue to exist because the concept of charity and not social justice is paramount. They are the salve for the community's conscience and a defense against drastic changes in the social order. Statutory agencies, whether public or voluntary, also subscribe to a similar ideology.

Bringing together agency and policy is fascinating grist for the mill of sociological analysis. The interactions of agencies; their power relations, and the implications for freedom and reason is not a well traversed territory. Hopefully, I have indicated the need for further study of community power from the frame of reference of an interrelationship of imperatively coordinated organizations.

For the urban dweller, the significant question is perhaps not should there be private or public agencies, but how is it possible to achieve social justice with an optimal balance of freedom and equality? The answer lies not in the inadequacies of present welfare structures per se but rather in the relations among organizations fostered by an ideology genuinely concerned with achieving welfare that embodies equality, freedom, and reason. Social systems of economic inequality can be changed by people taking the power from corporate influentials unto themselves.

APPENDICES

NOTES

REFERENCES

Appendix A

List of Abbreviations

Used in This Book

ACTION	Action Committee to Improve Opportunities for Negroes
AC	Arts and Education Council
ADC	Aid to Dependent Children
AFL-CIO	American Federation of Labor and Congress of Industrial Organizations
BCIC	Bicentennial Civic Improvement Corporation
BYU	Business, Young Men's Christian Association, and Urban League
Cath Ch	Catholic Charities
CC	Chamber of Commerce
CCRC	Civic Center Redevelopment Corporation
CEO	Chief Executive Officer
CEP	Concentrated Employment Program
CORE	Committee on Racial Equality
EWGCC	East-West Gateway Coordinating Council
HDC	Human Development Corporation
HEW	Department of Health, Education and Welfare
HPC	Hospital Planning Commission
HUD	Department of Housing and Urban Development
HWC	Health and Welfare Council of Metropolitan St. Louis
JCCA	Jewish Community Centers Association
JEVS	Jewish Employment and Vocational Service
JF	Jewish Federation

JOBS	Job Opportunities in the Business Sector
KSG	Knight of St. Gregory
LBJ	Lyndon Baines Johnson
MCF	Metropolitan Church Federation
MYC	Metropolitan Youth Commission
NAB	National Alliance of Businessmen
NAACP	National Association for Advancement of Colored People
OEO	Office of Economic Opportunity
RIDC	Regional Industrial Development Corporation
SCLC	Southern Christian Leadership Conference
UF	United Fund
USPHS	United States Public Health Service
WOU	Work Opportunities Unlimited
YMCA	Young Men's Christian Association
YWCA	Young Women's Christian Association

Appendix B

Highlights of Main Interviews

Both direct quotations and paraphrasing of interview responses are given. The question is stated and then the answer is included in quotes only if it is a direct quotation. Notes were taken during the course of each interview. They were written up afterwards. An effort was made to record what was said in the respondent's own words. The paraphrasing includes a summary of the respondent's remarks; explanations are enclosed in parentheses.

The interviews were informal with certain key questions included. These were varied depending upon the person being interviewed. These were in part focused interviews since certain questions were asked to clarify areas where my experience, or archives, newspapers, and local reports were insufficient for purposes of establishing the facts or making proper inferences.

The interviews are supplementary to the main line of data collection in confirming or disconfirming theories about agency relationships.

As case studies directed toward hypothesis formation and elaboration, this relatively unstructured, but focused, form of interviewing seems productive of spontaneous and insightful responses.

INTERVIEW 1

ROBERT K. SANFORD (*Post-Dispatch* staff member and author of the series of articles on Civic Progress, Inc.: interviewed January 5, 1967, concerning additional information on Civic Progress).

Question: What does Civic Progress think about using public money?

Answer: "It depends on the project . . . they are interested in things they can control. They have government contracts, and are used to doing business with the government. . . . They would not want to be known as against federal money. They really consider each thing separately."

Question: What is their major concern now?

Answer: "What has been done by Civic Progress is essentially brick and mortar. Much of this is completed or is on the way. The major issue now is social. Civil rights is seen as the number one problem. They see social services being taken care of by the government, and their contribution can be in the area of employment. They are bending some from the straight business viewpoint by trying to give some of those who are hired a chance to grow up on the job. If they don't make it on one job, perhaps they can on another."

Question: Why did Civic Progress decide to stay small?

Answer: "They were originally going to be a large organization with a staff. Instead, they decided to stay small and to work through existing organizations or set up new ones. . . . They could not get the executive they wanted. Also, they observed that some big community organizations did much talking but not much acting."

Question: Why was Downtown, Inc. set up?

Answer: "Aloys Kaufmann did not think that they could get the support of the Chamber of Commerce membership which includes many businessmen who have no downtown interests."

Question: What about the Regional Industrial Development Corporation?

Answer: "This came about because of the conservative influence of the Chamber of Commerce. Some retailers could be opposed to bringing competitors into the area."

Question: What is emeritus membership in Civic Progress?

Answer: "This is a nonvoting membership. Some of these guys feel that they have more to offer than the ones who are currently active." (He gave examples of Edwin Clark, previously chief executive officer of Southwestern Bell Telephone Co., and Ethan A. Shepley, previously Chancellor of Washington University and a former gubernatorial candidate.)

Question: How powerful is Civic Progress?

Answer: "It is still a force. There is so much money represented by those corporations. The banking interests in particular are closely tied together here and with Kansas City. There is a close tie in with the Kempers and Cravens." (The Kemper family dominates the banking community in Kansas City, Missouri, and Cravens is the head of Mercantile Trust Co. of St. Louis.)

Question: Why isn't labor in Civic Progress?

Answer: "Harrington (President and Board Chairman of Boatmen's National Bank) told me that they didn't know who could speak for or how much labor's viewpoint was needed."

Question: How powerful is labor?

Answer: "Morton May tried to get the Arts and Education Council in the UF, but some of the business people in the UF thought that labor would never buy it. . . . The success of the UF depends on labor's support."

Question: What about Civic Progress' relation to city and county governing bodies?

Answer: "They throw a party for the Board of Aldermen at the end of their session each year. They do the same thing for the County Council."

Question: Why isn't J. S. McDonnell (chief executive officer of Mc-Donnell-Douglas) in Civic Progress?

Answer: "He was invited, but he sent a vice-president, and they told him to forget it. They have his brother anyway." (William A. McDonnell, chairman of the finance committee of the St. Louis-San Francisco Railroad.)

Question: Any conflict between Civic Progress and the St. Louis Ambassadors? (These were appointed by the Mayor to help promote St. Louis.)

Answer: "Hell no. You don't name a building after the guys who only put on the top floor."

INTERVIEW 2

DAVID R. CALHOUN (Chairman of Civic Progress and President and Board Chairman of the St. Louis Union Trust Co.: interviewed March 9, 1967, concerning what is happening to voluntary agencies).

Question: What is happening to voluntary agencies given the power of public funds?

Answer: "Agencies may be divided into two groups: those which receive public funds and those which do not. Everybody is trying to get these federal grants and the government is stimulating it. This intrusion of the federal government is a political thing. These federal funds keep pouring in—nobody has got the guts to stand up and say no to these things. Conditions have gotten to the point in the city where there just doesn't seem to be any other way. The trouble is that people are in a hurry today—they can't wait to make it. It's part of the socialistic trend in the country. Besides the cost of the thing, it tells people that they'll do things for them that they can never do—such as eliminate poverty. You don't know what this will do to ignorant people—how they will react. There's the Communist influence out in Berkeley, and I just don't know how much that had to do with the Watts thing. The trend is for more and more government intrusion, and I don't think it can be stopped. . . . What was done for the Herbert Hoover Boys' Club is different. Somebody suggested that public funds could be obtained to buy something that was needed. I said: 'God damn, we've done the whole thing with private money up to this point—let's keep it that way.' "

Question: Why have the UF allocations increased only three percent per annum since 1960?

Answer: "It isn't enough is it? Some of the agencies are so expensive to operate. The per capita costs of some day care centers are high. There is also the matter of making the goals. Sometimes we set them too low, I realize. But it is so important to make the goals, and for the image of the community. . . . It is a helluva way to raise money for the agencies and institutions. You can't beat it in terms of the cost of raising the dough. . . . Most give willingly but you have to twist a few arms."

Question: What about using UF money for matching funds for HDC programs, even if matching requirements go up to twenty percent—will we let the federal money go?

Answer: "We won't do that. It is a helluva good deal—it's too good to turn down."

INTERVIEW 3

JOHN MYRICK (former Campaign Manager of the United Fund: interviewed January 6, 1967).

Question: What about all the campaigns in the community?

Answer: "They are of concern to the executive committee and agency relations cabinet. However, federation is a voluntary matter. These drives continue because people continue to work in them and give to them. The Easter Seals is a very expensive campaign in terms of the returns on what is sent out. These health outfits get very little more than token contributions from the big corporations."

Question: What do the business leaders think about the War on Poverty?

Answer: "These are practical men. Even though they are conservative, they will support something when they know there is nothing which they can do about it. This is the way they run their own businesses. If they see that something is losing money for them—or they could make more money by closing out a product line—they will act. The same way with the hospitals. They would prefer to continue their support, but since the government has gotten into this, they are simply forced to get out because it is just good business sense. They figure that some one needs to be in there to see what is going on."

INTERVIEW 4

WILLIAM KAHN (Executive Director of the Jewish Community Centers Association: interviewed January 3, 1967).

Question: What is the future of voluntary agencies?

Answer: "We can expect more governmental funding of all kinds of welfare programs. . . . But the needs are great, and there is still much room for voluntary agencies to be innovators. Also there is a need for an outlet for philanthropy. The psychological reward in private philanthropy is still great. It is not weakening—fund raising is stronger than ever. Multiple fund drives give many people a sense of goodness, humaneness. . . . The Chamber and Fund may be concerned, but on the whole this is a healthy sign of community life. The Jewish community has a hundred drives going on. . . . There is some thinking that when the "fed" money comes in, the private money should be pulled out."

Question: Any difference in political outlook between affluent Jews and gentiles?

Answer: "The affluent Jews are very conservative—no different than the gentile affluent in worrying about the free enterprise system."

Question: What about control that may come with accepting public funds?

Answer: Mr. Kahn indicated that he is not troubled with controls that can come with receiving federal money. Federal funds are needed for special services which could not be financed except through public money. "There is a lot of red tape and control with the private dough also. There have been few strings with the federal money that the Center (JCCA) has had thus far. There are a lot of stereotypes on the business of taking fed money."

Question: Does JCCA have the top leadership of the Jewish community?

Answer: Kahn mentioned JCCA has a Board of Trustees and Irving Edison, in Civic Progress, is on that board. "They are the board behind the board . . . an informal group of key guys who will backstop when necessary."

Question: What about the Jewish Federation?

Answer: "It has not been as successful here as in other communities—not the commitment of the corporate elite among Jewry. . . ."

Question: How powerful is Civic Progress?

Answer: He says that Civic Progress has considerable influence in the community. "When JCCA needed to get contributions from the gentile community, Russell Dearmont got some of the guys together who kicked in what was asked."

INTERVIEW 5

WILLIAM DOUTHITT (Executive Director of the St. Louis Urban League: interviewed January 10, 1967).

(He was in the midst of a capital funds drive to pay for the building into which the Urban League had moved. He said that it was "tough going." He mentioned working with Wesley McAfee, Chief Executive Officer of Union Electric, and G. S. Roudebush, Vice President and General Counsel for McDonnell Douglas.)

Question: What about Civic Progress?

Answer: "These men have power because they control the money in the community . . . They will let you do pretty much what you want in the way of social welfare as long as you don't interfere with their money making."

Question: How is Work Opportunities Unlimited doing?

Answer: Mr. Douthitt described it as a "protective device" and that "management criteria" are being used to run it. It is competing with the Urban League.

Question: How is citizen participation in the antipoverty programs working out?

Answer: "Maximum feasible participation of the poor is a pain in the ass. . . . These people try to take over the program. They want to be given the money to spend for themselves. . . . When I tell them that they do not have the knowledge to run these things, they call me a Caesar."

Question: What about Gibbons' influence? (Harold J. Gibbons is head of the Teamsters Joint Council.)

Answer: "He is a force politically . . . a career diplomat . . . who has made a kind of accommodation with business. . . . They are getting no place trying to organize Famous-Barr and did you see how quickly Egan stepped down from the firefighters when they went on strike? (Egan was the head of the local and also a full-time paid labor representative for

the United Fund.) They all get together for good of St. Louis . . . pavilion, stadium and things like that."

Question: What is the Model City Agency going to do?

Answer: "This is a political boondoggle. HUD is the force. . . . It
(Model City Agency) will be handled through the Mayor's Office. Social
workers will have little to say . . . look at the vested interests involved."

Question: What about McDonnell Douglas and Union Electric's involvement in the Model City Agency?

Answer: "Sure, because they see this as an opportunity to make
money."

Question: What should be the role of the voluntary agencies?

Answer: Douthitt answered by saying that he is pushing for a more
forceful front in the voluntary sector. "Now is the time to come up with a
program for the private agencies. Many of these public programs are
going to fall on their ass. . . . The private agencies should make a proposal to the foundations that we could do this for one-fourth less money."

INTERVIEW 6

WALTER WAGNER (Executive Director of the Metropolitan
Church Federation: interviewed September 18, 1967).

Question: Who really controls social welfare in St. Louis?

Answer: "The Health and Welfare Council . . . it is the right arm
of the United Fund—it is a validating organization. The Council determines
what programs will be included, and what agencies will be recommended
to the United Fund. . . . We have been turned down in our applications
for acceptance as a UF member. . . . A past president of the UF told
me—'You let me know when you want to try again, and I will go with you.'
. . . The chances are better for getting in because of the flack coming
from the churches over all the Catholics are getting from the UF . . .
HWC has a naive definition of social work. They view church social work
as substandard. If this were true, I would agree that we should not be in
the UF. . . . They look upon Jewish and Catholic social work as being
of higher standard . . . What determines the allocations from the United
Fund is who gets on the board. The whole thing is a power structure. You
get on the Downtown Y, and anybody who is somebody graduates to the
Metropolitan board. From there you move to the Boy Scout board, which
shows that you have come to the top of the ladder. It is a steady kind of
graduation. I have not played the game of power politics—I've refused to
use it."

Question: What do you think about direct federal grants to churches?

Answer: "I am opposed. The temptation for churches to use such funds on a wide nonsectarian basis might be too great to handle. It is contrary to the First Amendment of the Constitution. The church needs to be able to criticize and not lose its voice."

INTERVIEW 7

REV. JOSEPH A. MCNICHOLAS (Secretary of Catholic Charities: interviewed September 8, 1967).

Question: What is the relationship of Catholic Charities to the United Fund?

Answer: He said that Catholic Charities does no fund raising or budgeting for the agencies. Basically, they are concerned with standards. Budgets go through the administrative office of Catholic Charities. They do, he said, exercise some control to see that the budgets are in line with UF requirements.

Question: What has been Catholic Charities' connection with the War on Poverty?

Answer: "Because Catholic Charities has not gotten into the poverty programs, they say we are a member of the 'establishment.' We have a grant of $363,000 from the United Fund and must maintain the confidence of the UF that we are doing a good job. Some priests are critical because they lack the funds to engage in activities which they believe are needed. . . . If someone else can take over the work of Catholic Charities, then we would gladly go into some other field of endeavor. . . . Catholic Charities could not maintain its professional standards if it got into some of the poverty programs. . . . We were cited by the Child Welfare League because we have not diluted our services. . . . The Monfort Brothers tried to get the Juvenile Court to release children to their custody, but these priests do not have the professional background to deal with these kinds of problems. . . . On the other side, you will hear that we kept them from getting money to help delinquent children. . . . I never saw anything that was really successful that was not well organized."

Question: What will happen to sectarian welfare with the increased use of public funds?

Answer: He answered that he envisions no substantial change in Catholic Charities or sectarian welfare because of the greater availability of public funding. "The main thing is a first class service, and if the State of Missouri can do a better job, then we'll think about doing something else. . . . The protection of religious interests are generally accepted and the state can do this too. We already have Catholic children under public care. . . . We are suffering by all this talk about public funds. This is reflected in higher circles of the Church. I favored action that would upgrade services at Father Dunne's Home for Children. The Cardinal said, 'You don't want to get involved in that. Let the State take care of the matter.' After Vatican II the Bishops are oriented to using public funds. But there is such an inadequacy of public money for many services that private funding will be required for a long time to meet needs."

Question: Does the Protestant Establishment run things in St. Louis?

Answer: "When I give talks to Catholic groups about supporting the United Fund, some will ask me if it is a Protestant agency, and I tell them you are going to have to fight me on this. I then point out the representation which Catholics have in the United Fund." He commented that although Catholics have not penetrated upper levels of the business community as much as Protestants, the former are politically dominant in the City of St. Louis.

Question: What accounts for the clustering of board members of Catholic Charities in the central west portions of the city?

Answer: He explained that they want prestigious people who can command community-wide respect.

INTERVIEW 8

ANTHONY DeMARINIS (Executive Director of Family and Children's Service: interviewed December 20, 1966).

Question: What is the future for voluntary agencies?

Answer: "The voluntary field is stronger than ever. We have had this talk since the 1930s about the voluntary agencies going out of business. The voluntary agency has existed in spite of the public agency." He referred to a speech by Arthur Kruse ("The Future of Voluntary Welfare Services," address at Annual Meeting of Community Fund of Chicago, January 21, 1965:9) which indicates the magnitude of the public programs as compared to the recent ones funded by the Economic Opportunity Act. DeMarinis commented on Levin's (1966:102) article about how the roots of voluntary agencies stem from fears of religious and political tyranny, and that the First Amendment to the Constitution guarantees freedom of assembly, separation of church and state.

Question: What are the possibilities for partnership with the public agency?

Answer: "Partnership means that you have to serve who they tell you to serve. If public funds are accepted by the voluntary agency, it should be on the voluntary agency's terms. It should be like when McDonnell Aircraft accepts a contract from the government. There is a definite terminal date, and it's under the control of the company to deliver on its terms and specifications. Agencies should not be pushed to take large chunks of public money. The UF is greater than the sum of its parts . . . should not try to hammer out an average agency—need to have variety. . . . The War on Poverty is holding back the development of the public agency. With all this hullabaloo, attacking both the public and voluntary agencies, OEO is keeping the idea of public services from being extended. For example, we should be pushing homemaker services for the public agency—we keep doing it so that it stays in the voluntary field."

Question: What should the voluntary agencies be doing?

Answer: "There is a big service area for the voluntary agencies. About 80 percent of the people are not in poverty." He quoted from Kruse: "From 1950–1962, the percentage of families earning $10,000 and over increased from 3.2 percent to 17.7 percent of the population. Families earning over $6,000 increased from 14.2 percent to 49.6 percent. Those earning less than $3,000 declined from 42.6 percent to 19.9 percent. The growing productivity of our society suggests that ten years from now a hard core of 8–10 percent severely disadvantaged economically will still exist, but the median income will have risen from the present $5,940 to some $8,000. The proportion of families earning $10,000 and over could well increase from 17.7 percent to 25–30 percent. Families earning over $6,000, over the next ten years, might increase from 49.6 percent to 75 percent of the population. More people will be able to finance their own solutions to social problems without being the full beneficiaries of either tax funds or contributed dollars. The fact remains, and this is my second conclusion, that this growing middle and upper middle class will continue to be plagued with the social problems which are the inevitable by-product of an artificial society." DeMarinis continued by saying: "There are so many needs not being met that because of the costs, we will not have enough public services for a long time. There is an increased need to work with individuals in the business setting on matters such as absenteeism, alcoholism and divorce. . . . We need to take a hard look at the contributed dollars. The increase will not be great from this source, and we must be sure that it is well spent. . . . Also, we should look more to fees for service. . . . The emphasis should be on prevention rather than on treatment. . . . Don't overlook the growth of the mental health concept. The family life education programs of the agency show the same kinds of questions regarding child rearing in parent education groups whether they are in Pruitt-Igoe or in west county. The need for community psychiatry is great."

INTERVIEW 9

RAYMOND F. TUCKER (former Mayor of St. Louis: interviewed March 7, 1967).

Question: Who controls social policy in St. Louis?

Answer: "I would like to know the answer to that one too. I don't know that anyone does. You know there was the Social Planning Council (now the Health and Welfare Council) which was supposed to do this sort of thing. But they never did it. They never could do anything about all the agencies . . . the duplication of services. I became very much concerned about delinquency, and was responsible for the formation of the Youth Commission. The Council was not able to do anything in this field so we set this up. Putzell (Edwin J. Putzell, Jr., Secretary of Monsanto) asked me about the Council and we had lunch one day and I told him quite frankly that the Council wasn't doing the community planning job that it should be doing. (Putzell was President of the Council at the time.) Rudy (Rudolph T. Danstedt, former Executive Director of the Council, when Putzell was its President) was a nice guy to talk to, but he couldn't get things done. They reorganized the Council after Rudy left, changed the name, but as far as I could tell they weren't doing anything different than before. They really had a chance under my administration to be something—to do the kind of job of pulling these health and welfare services together, but they never took advantage of it." He told about setting up the Juvenile Delinquency Control Project, and how it was the first in the country to qualify for funds. This was more than a delinquency project, he commented, because it was really to provide centralized coordinated planning for various health, welfare and recreation projects. He said that he merged the Concerted Services Task Force (organized with the assistance of the Health Education and Welfare Department to coordinate services to the Pruitt-Igoe Housing Project) and the Juvenile Delinquency Control Project. It was to bring together, he related, both private and public agencies in the interests of community welfare.

Question: What about the different community agencies receiving federal funds?

Answer: He said that he was opposed to a lot of these federal grants because they are not really coordinated or tied into sound planning. Consequently, he said, these grants are delaying the institution of real planning of substance.

Question: What about the role of Civic Progress, Inc. in St. Louis?

Answer: "St. Louis had not had any bond improvements for some years and as a result, the city was in bad shape—one-third of the residential properties were deteriorating or dilapidated and the city was not able to meet expenses." He spoke of the efforts of Civic Progress in obtaining the support of the Missouri State Representatives in approving the Earnings Tax for the City of St. Louis. They went out to the legislators' towns and had "blue plate specials" to talk them into backing the tax plan. Concerning the bond issues he said: "One of the group would agree to head the drive, draw up a budget and submit it to Civic Progress, and the members would then put up the money. . . . They never allowed any substitutes. They wanted people to be on the same level as themselves, who could make the decisions without having to run back to talk to someone else. When they would run into problems with some of the boys, they could call a member of Civic Progress and explain the situation to him. Then he would hear from a Civic Progress member who would say that 'so and so understands, now you won't have any trouble in getting his cooperation.' There were never any strong-armed tactics—just explaining the situation."

Question: What do you see is the role of the East-West Gateway Coordinating Council?

Answer: "This is the beginning of comprehensive planning. . . . I would like to see the Gateway Council big enough and representative of the various civic leaders. . . ."

Question: How strong is the Mayor's Office?

Answer: "It is not strong at all—it is only strong because people think it is." He spoke of how few patronage jobs there are, and the Police and Board of Education being separate from the city.

Question: Why do the aldermen usually go along with the Mayor?

Answer: "Most issues are in the interest of the community. They are the kinds of matters which would be difficult to disagree with if you are interested in making St. Louis better."

Question: What is the future of the private agencies?

Answer: "All money should come through a central corporation with no grants to private agencies. The staff of the private agencies could be hired by the central organization. Now there is too much overlapping of agency programs. A lot of people are getting money and don't know whether it is effective or not."

INTERVIEW 10

Eugene C. Moody (Executive Director of East-West Gateway Coordinating Council: interviewed October 17, 1967).

Question: What are the Council's relations with the large corporations?
Answer: "We are so young that there have been relatively little. We

have had rather intensive communication with McDonnell, but it has been transportation oriented. . . . It's got to be improved. . . . We want to get a business orientation and need strong participation of this kind regarding regional studies, land use, and matters of a physical nature. We've had a close relationship with RIDC—both board and staff, but we have not done this as much as we would like to. It's essential if we are going to get these programs implemented. Without the power structure's support, or their acquiescence, we can't get things done. . . . In any operation I think of their help coming in many ways. With McDonnell it has been in terms of moving people and goods—pure research—development of optimum ways to meet goals of transportation and choice. . . . It is not political implementation alone. We need this peculiar knowledge and their influence."

Question: Any structural way for business to be tied into the Council?

Answer: "There is no structural way. At some point in our career we need a citizen's board, quarterly presentations to Civic Progress or some other way."

Question: Has the Council made any presentations to Civic Progress?

Answer: "We've had an indirect presentation regarding a study of rapid transit. Civic Progress intermediaries or investigators came to us and we spent a day discussing a viable plan."

INTERVIEW 11

SAMUEL BERNSTEIN (Executive Director of the Human Development Corporation: interviewed December 27, 1966).

Question: What is happening about the use of public funds by the voluntary agencies?

Answer: "I see it the same way as I discussed in my paper ("Public Funds and the Private Agency," presented to the United Fund Agency Executive Director's Meeting at Camp Beaumont, May 7, 1965). We still don't have a commitment from the UF regarding financing of poverty programs. The feeling is that because government is in it we should not be in. In the voluntary field there is no redirection of funding—no new programs and the increases are minor. There is no overall commitment to how we can couple the poverty programs with private agency programs. They want to take the OEO funds and call the shots. . . . They are afraid of the costs."

Question: What about Work Opportunities Unlimited?

Answer: "One or two individuals put up $10,000. . . . Ethan Shepley made some commitment, but this is atypical here. For the most part they have boycotted the War on Poverty. They have a business point of view on jobs . . . here is a job and the applicant is either qualified or not. WOU up to this point is a token effort."

Question: What about the Health and Welfare Council?

Answer: "It is an arm of the Fund . . . takes directions from the UF. Say HWC-UF in the same breath. There is recognition that the professional staff wants to do other things, but as long as funding comes from the Fund, they are inhibited."

INTERVIEW 12

HOWARD BUCHBINDER (formerly a community organizer for the Human Development Corporation and now at the St. Louis University School of Social Service: interviewed March 13, 1967).

Question: What can you tell me about political influences on the Human Development Corporation?

Answer: "Political influences concerning HDC are more indirect than direct. They are more subtle than the political domination that has taken place in Chicago where Shriver was sent away with his tail between his legs." Buchbinder described the process of appointment from the Neighborhood Advisory Committees to the Citizen Advisory Council and to the Board of the Human Development Corporation which involves approval of the Mayor. "If Bernstein is doing his job, he will sit down with the Mayor and go over the list of names so that they pick those who are least likely to give trouble." He referred to a letter from Father Lucius Cervantes that indicated the importance of consensus and rational approaches to the solution of community problems. Buchbinder's interpretation was that Father Cervantes is trying to avoid any kind of militancy. Buchbinder spoke of the Conference of Mayors where Cervantes said that he wanted to have more say-so about the antipoverty programs in St. Louis. Also, Buchbinder mentioned that the Governor's approval is needed on various programs, and that he can veto proposals. . . . The Model City Agency, Buchbinder stated, is political and referred to the use of HDC staff in working up the proposal for the Model City Agency . . . Rather than coming from the politicians per se, he said, the pressure has come from the "downtown staff who have cautioned against moving too fast or taking too hard a line. When the going got risky, they suggested that they could handle it. This is a paternalistic point of view and undercut efforts to build an effective form of protest among the poor." Buchbinder said it is often the middle class who really represent the neighborhood. This is true of those on the Citizen Advisory Council and the board of HDC. Although these are people who live in the neighborhood, it made for conservatism, according to Buchbinder, in matters regarding protest. "I could have gone along with the idea of compromise—'go slow'—'this is OK but not at this time'—if the commitment of the downtown staff had been to really change conditions. . . . But this was not so. Instead, they wanted to give services to people; i.e., the Galbraith case approach to poverty." HDC programs, Buchbinder said, departed from the basic mission of the War on Poverty. "The Economic Opportunity Act provided for radical innovation —the participation of the poor in changing their conditions." He referred to a speech he gave ("In Poverty—Who Speaks for the Poor," presented at Berea Presbyterian Church, February 12, 1967:4) where he made this same point. "It denied that the poor are at fault. . . . Poverty is not something that just happens. There are groups in the society that make a profit from the poor. There are groups whose profits or power would be threatened if the poor would be secure economically and active social and political participants. . . ." Specifically, in this speech he criticizes the poverty programs because they are basically not directed toward institutional changes. The Human Development Corporation, Buchbinder said, is interested in safe programming. He disagrees with Bernstein (May, 1965 speech cited in interview 11) that the fundamental idea of War on Poverty is to give people a chance to help themselves. Buchbinder said in his speech: "The underlying purpose of the War on Poverty is to place funds at the disposal of the impoverished community of America to wage war against the cause of poverty—not to build a mammoth service program to develop human beings as Mr. Bernstein states."

Question: What would be an example of this kind of paternalism?

Answer: This is evidenced, he says, in what is done for them by the employment services. He thinks more would have been accomplished if people had been allowed to carry through with the protest which had been started because there were no employment counselors in the neighborhood stations. While he was out of town, about thirty people went down to the Washington Ave. office asking to see DeLargy and Fogler (officials in the St. Louis Office of the Missouri Employment Service). The guard, Buchbinder said, became excited, and the local office called Julian in Jefferson City, the head of the Missouri State Employment Service, who in turn called the Governor who held up signing the Comprehensive Manpower Program Proposal. HDC's central office was disconcerted, Buchbinder said. They contended that his people had jeopardized the whole Comprehensive Manpower Program, and a number of agencies could be hurt in the process besides HDC. They advised him to persuade his people to hold off, because if the Governor approved, there would be a decentralization of staff anyway. Buchbinder agreed to explain what had happened and asked them to wait. He said: "I believe that I made a mistake in doing so."

NOTES

1. For a discussion of public ideology see pp. 30–31.

2. Archives of Morris Keeton's Office, Academic Vice-President, Antioch College.

3. See Pifer (1966). Definition includes private and nonprofit agencies, but arbitrarily excludes universities, hospitals, or religious missions for his purposes, he says in this article.

4. *The Fortune Directory, Fortune,* July, 1963, 1964, 1965, 1966 and June 15, 1967. Excluded are Wagner Electric, Universal Match, and General Steel Industries which were included in 1963 but dropped out during this five year period. This includes employees in and outside of St. Louis area.

5. The Health and Welfare Council makes a study every five years of expenditures for social and health services in the United Fund area. This study is under the auspices of United Community Funds and Councils, and follows restrictive definitions of what should be included as social welfare. However, many of the services excluded, although fitting social policy in the broader sense of the Welfare State, are actually settled social policy: for example, some Social Security programs. In effect, they have been already decided by the corporations—they are a part of the juridical structure of St. Louis and the nation.

6. Estimated for 1966 by the Jewish Federation of St. Louis.

7. Special tabulation of data obtained from the 1965 Expenditure Study, conducted by the Health and Welfare Council of Metropolitan St. Louis. This is unofficial because the Council does not report allocations separately by the major religious bodies. Jewish agencies belonging to the United Fund are: Jewish Hospital, Jewish Community Centers Association, Jewish Center for the Aged, Jewish Employment and Vocational Service, Jewish Family and Children's Service, and the Jewish Federation.

8. See Figure 2 and Tables 15, 17, and 21 for more detail. These amounts do not include the costs of programs which are considered part of a church's program of religious training.

9. For other information on Civic Progress, see Chapters 3–7.

10. *Gateways for Youth,* St. Louis: St. Louis Human Development Corporation, 1964, pp. 210–214. This gives background information on the origins and development of delinquency planning in St. Louis.

11. "A Dream Deferred: The Story of Pruitt-Igoe's Conditions," n.d. Also see "A Preliminary Report on Housing and Community Experiences of Pruitt-Igoe Residents," Pruitt-Igoe Project Social Science Institute, Washington University, St. Louis, Missouri, May 13, 1966.

12. Reverend Lucius Cervantes shuttles back and forth, on special assign-

ments, between the Mayor's Office and the sociology department of St. Louis University.

13. "Mayor Predicts City to Gain in Population," *St. Louis Globe-Democrat* (July 22, 1966). This came as a response to an estimate made by the United States Census Bureau that the City's population had dropped to the lowest level since 1910.

14. Lipsky (1967:22–28) for discussion of tactics used to protect target groups.

REFERENCES

AGGER, R. E., D. GOODRICH and B. E. SWANSON (1964) *The Rulers and the Ruled.* New York: John Wiley.

ALMOND, G. A. and S. VERBA (1963) *The Civic Culture: Political Attitudes and Democracy in Five Nations.* Princeton: Princeton University Press.

BAINBRIDGE, J. (1962) *The Super Americans.* New York: Random House.

BALTZELL, E. D. (1966) *The Protestant Establishment, Aristocracy and Caste in America.* New York: Random House.

BANFIELD, E. C. (1965) *Big City Politics, A Comparative Guide to the Political Systems of Nine American Cities.* New York: Random House.

BANFIELD, E. C. and J. Q. WILSON (1966) *City Politics.* New York: Vintage Books.

BELL, D. (1960) *The End of Ideology.* Glencoe: The Free Press.

BERELSON, B. and G. A. STEINER (1965) *Human Behavior: An Inventory of Scientific Findings.* Chicago: Harcourt, Brace & World.

BERNSTEIN, S. (1965) "Public funds and the private agency." Presented to the United Fund Agency Executive Meeting at Camp Beaumont, May 7.

BLAU, P. M. (1955) *The Dynamics of Bureaucracy.* Chicago: University of Chicago Press.

BLAU, P. M. and W. R. SCOTT (1962) *Formal Organizations: A Comparative Approach.* San Francisco: Chandler Publishing.

BLOOMBERG, W. and H. J. SCHMANDT, Eds. (1968) *Power, Poverty and Urban Policy.* Beverly Hills, Calif.: Sage Publications.

BOLLENS, J. C., Ed. (1961) *Exploring the Metropolitan Community.* Berkeley and Los Angeles: University of California Press.

BOLLENS, J. C. and H. J. SCHMANDT (1965) *The Metropolis, Its People, Politics and Economic Life.* New York: Harper & Row.

BOWERS, R., Ed. (1966) *Studies on Behavior in Organizations.* Athens: University of Georgia Press.

BRODINE, V. (1964) "The strange case of the Jefferson Bank vs. CORE." *Focus Midwest* 2, No. 10:14.

BURNHAM, J. (1941) *The Managerial Revolution.* New York: John Day.

CALLOWAY, E. (1965) "St. Louis mayoralty elections: a reform period closes." *Focus Midwest* 3:14.

——— (1964) "The nature and flow of economic power." *Missouri Teamster* 1, No. 11.

CATHOLIC CHARITIES OF ST. LOUIS (1966) *Annual Report.*

CERVANTES, A. (1967) "To prevent a chain of super-Watts." *Harvard Business Review* (September–October): 55–64.

CHAMBER OF COMMERCE OF METROPOLITAN ST. LOUIS (1966a) *Metropolitan St. Louis Large Employers, 1966.*

——— (1966b) *Guide to Giving.* Prepared by the Charities Committee of the St. Louis Chamber of Commerce, September.

CHINOY, E. (1961) *Society: An Introduction to Sociology.* New York: Random House.

COHEN, N. E. (1964) "Future welfare policy, program and structure." Pp. 3–19 in *The Social Welfare Forum, National Conference of Social Welfare.* New York: Columbia University Press.

DAHL, R. A. (1961) *Who Governs? Democracy and Power in an American City.* New Haven and London: Yale University Press.

DELUGACH, A. (1968) "Businessmen are vital to Negro job programs." *St. Louis Globe-Democrat* (February 26).

DIVISION OF WELFARE (1964) *The ADC Families of Pruitt-Igoe, A Descriptive Study, 1962.* St. Louis: Missouri Department of Public Health and Welfare (February).

DOMHOFF, G. W. (1967) *Who Rules America?* Englewood Cliffs, New Jersey: Prentice-Hall.

EDELSTON, H. C. (1964) "Can communities still plan from the local community point of view?" Proceedings of the Fifteenth Annual Adirondack Workshop, National Social Welfare Association, Inc., July 2–8.

ELMAN, R. M. (1966) *The Poorhouse State.* New York: Random House.

ETZIONI, A. (1964) *Modern Organizations.* Englewood Cliffs, New Jersey: Prentice-Hall.

——— (1961) *Complex Organizations: A Sociological Reader.* New York: Holt, Rinehart & Winston.

FORTUNE MAGAZINE (1963, 1965, 1966, 1967) *The Fortune Directory* (July all years but 1967, in which June 15).

——— (1967) "The road to 1977." *Fortune* (January): 93ff.

GOULDNER, A. G. (1965) "Organizational analysis." Pp. 400–428 in R. K. Merton, L. Broom and L. S. Cottrell, Jr., Eds. *Sociology Today,* Vol. 2. New York: Harper & Row.

——— (1952) *Patterns of Industrial Bureaucracy.* Glencoe: The Free Press.

GOULDNER, A. W., D. J. PITTMAN, L. RAINWATER and J. S. STROMBERG (1966) *A Preliminary Report on Housing and Community Experiences of Pruitt-Igoe Residents.* St. Louis: Pruitt-Igoe Project, Social Science Institute, Washington University (May).

GREER, S. (1963) *Metropolitics: A Study of Political Culture.* New York: John Wiley.

GUETZKOW, H. (1966) "Relations among organizations." In R. V. Bowers, Ed. *Studies on Behavior in Organizations.* Athens: University of Georgia Press.

GUNN, S. and P. S. PLATT (1945) *Voluntary Agencies.* New York: Ronald Press.

HACKER, A., Ed. (1965) *The Corporation Take-Over.* Garden City: Doubleday.

HAMLIN, R. H. (1961) *Voluntary Health and Welfare Agencies in the United States.* New York: Schoolmasters Press.

HARRINGTON, A. (1959) *Life in the Crystal Palace.* New York: Alfred A. Knopf.

HARRINGTON, M. (1963) *The Other America.* New York: Macmillan.

HAUSER, P. M. (1967) "Social goals as an aspect of planning." Prepared for discussion at the 1967 National Health Forum Planning for Health, Chicago, March 20–22.

HEALTH AND WELFARE COUNCIL OF METROPOLITAN ST. LOUIS (1967a) *Fifty-sixth Annual Report.*

———— (1967b) *Population by Census District St. Louis City.* St. Louis: HWC and Metropolitan Youth Commission (January).

———— (1966) *The St. Louis Cathedral Parish Area, A Study of People, Physical Conditions, Problems and Needs* (November 30).

———— (1965a) *Community Service Directory.*

———— (1965b) *Memorandum to Bicentennial Civic Improvement Corporation Review Committee* (August 5).

———— (1965c) Report No. 7 to Board of Directors (October 26).

HOROWITZ, I. L. (1967) "Social science and public policy: an examination of the political foundations of modern research." *International Studies Quarterly* 11 (March).

———— (1964a) "A formalization of the sociology of knowledge." *Behavioral Science* 9, No. 1 (January): 45–55.

————, Ed. (1964b) *The New Sociology.* New York: Oxford University Press.

————, Ed. (1963) *Power, Politics and People: The Collected Essays of C. Wright Mills.* New York: Oxford University Press.

HOROWITZ, I. L. and M. LIEBOWITZ (1968) "Social deviance and political marginality: toward a redefinition of the relation between sociology and politics." *Social Problems* 15 (Winter): 280–296.

HUNTER, F. (1959) *Top Leadership, USA.* Chapel Hill: University of North Carolina Press.

———— (1953) *Community Power Structure.* Garden City: Doubleday.

INFORMATION (1967) *For Members of Board of Directors of the United Fund of Greater St. Louis, Inc.*

JACOBS, R. (1966) "HDC pay increase coincided with Seay's CORE resignation." *St. Louis Post-Dispatch* (September 13).

JOHNS, R. E. (1946) *The Cooperative Process Among National Social Agencies.* New York: Association Press.

JOINER, C. A. (1964) *Organizational Analysis: Political, Sociological and Administrative Processes of Local Government.* Lansing: Michigan State University Press.

KRAVITZ, S. (1968) "The community action program in perspective." In W. Bloomberg, Jr. and H. J. Schmandt, Eds. *Power, Poverty and Urban Policy.* Beverly Hills, Calif.: Sage Publications.

LEVIN, H. (1966) "The essential voluntary agency." *Social Work* 11, No. 1 (January): 98–106.

LEVY, J. (1928) "Historical analysis of the evolution of the Community Fund of St. Louis." Unpub. Master's thesis, Department of Sociology, Washington University.

LINGLE, W. L. (1966) "The voluntary agency's mission in the great society." Address at the annual meeting at the Community Chest Council of the Cincinnati Area, March 30.

LIPPINCOTT, E. and E. AANNESTAD (1964) "Management of voluntary welfare agencies." *Harvard Business Review* (November–December): 87–98.

LIPSET, S. M., M. A. TROW and J. S. COLEMAN (1956) *Union Democracy.* Glencoe: The Free Press.

LIPSKY, M. (1967) "Protest as a political resource." Institute for Research on Poverty, University of Wisconsin (Discussion Papers).

LONG, N. E. (1962) *The Polity.* Chicago: Rand McNally.

———— (1958) "The local community as an ecology of games." *American Journal of Sociology* 64 (November): 257.

LUNA, M. (1967) "Yes! gateway council is getting the job done." *St. Louis Globe-Democrat* (March 11–12).

LUNDBERG, F. (1968) *The Rich and the Super-Rich.* New York: Lyle Stuart.

McNEIL, C. F. (1964 and 1965) "Community organization, its practice in United Funds, Community Chest and community health and welfare councils, and some of the factors affecting its course." Lecture delivered at Schools of Social Work of Boston College (Fall, 1964) and University of Michigan (Spring, 1965).

MANSER, G. (1965) "Voluntary organization for social welfare." In *Encyclopedia for Social Work.* New York: National Conference of Social Workers.

MANUAL OF ST. LOUIS BANK STOCKS (1966) St. Louis: G. H. Walker (March).

MARCUSE, H. (1966) *One-Dimensional Man.* Boston: Beacon Press.

MARRIS, P. and M. REIN (1967) *Dilemmas of Social Reform.* New York: Atherton Press.

MARSHALL, T. H. (1965) *Social Policy.* London: Hutchinson University Library.

MATSON, F. W. (1966) *The Broken Image.* New York: Anchor Books, Doubleday.

MAY, E. (1964) *The Wasted Americans.* New York: Harper & Row.

METROPOLITAN ST. LOUIS HOSPITAL PLANNING COMMISSION (1966) *Evaluation of a Capital Fund Project Proposed by St. Joseph Hospital.* Prepared by Metropolitan St. Louis HPC (April).

METROPOLITAN ST. LOUIS SURVEY (1957a) *Background for Action.* St. Louis: Metropolitan St. Louis Survey (February).

———— (1957b) *Path of Progress for Metropolitan St. Louis.* St. Louis: Metropolitan St. Louis Survey (August).

MICHELS, R. (1959) *Political Parties.* New York: Dover.

MILLER, S. M. (1964) "Poverty, race and politics." In I. L. Horowitz Ed. *The New Sociology.* New York: Oxford University Press.

MILLS, C. W. (1956) *The Power Elite.* New York: Oxford University Press.

MONSENS, R. J., JR. and M. W. CANNON (1965) *The Makers of Public Policy: American Power Groups and Their Ideologies.* New York: McGraw-Hill.

MOORE, W. E. (1962) *The Conduct of the Corporation.* New York: Random House.

MORGENTHAU, H. J. (1968) *Politics Among Nations.* New York: Alfred A. Knopf.

MOSCA, G. (1939) *The Ruling Class* (H. S. Kahn, translator). New York: McGraw-Hill.

NATIONAL ADVISORY COMMISSION ON CIVIL DISORDERS (1968) *Report of the National Advisory Commission on Civil Disorders.* New York: Bantam Books.

New York Times (1969a) "U.S. may oppose conglomerates in trust actions." (March 13): 1.

———— (1969b) "World stock ownership is urged." (March 8): 39.

OFFICE OF ECONOMIC OPPORTUNITY (1967) *Catalog of Federal Assistance Programs.* Office of Economic Opportunity (June 1).

OLDS, E. B. and W. H. SCHMIDT (1947) *St. Louis Looks at Its Community Chest.* St. Louis: Greater St. Louis Community Chest.

PACKARD, V. (1963) *The Waste Makers.* New York: Pocket Books.

PIFER, A. (1966) "The nongovernmental organization at bay." In the 1966 *Annual Report, Carnegie Corporation of New York.*

"Polk's St. Louis (Missouri) city directory 1966" (1967) Detroit: R. L. Polk.

PRESTHUS, R. (1965) *The Organizational Society.* New York: Random House.

———— (1964) *Men at the Top.* New York: Oxford University Press.

RAINWATER, L. (1967) "The services strategy vs. the income strategy." *Trans-Action* (October): 40–43.

RAINWATER, L. and W. L. YANCEY (1967) *The Moynihan Report and the Politics of Controversy.* Cambridge: M. I. T. Press.

"Religion and the great cities" (1967) Great Cities Series, St. Louis University (March 29).

RIDGEWAY, J. (1969) *The Closed Corporation.* New York: Ballantine Books.

ROSE, A. (1967) *The Power Structure.* New York: Oxford University Press.

ROSEN, S. (1939) "The historical development of the Jewish federation of Saint Louis." Unpub. Master's thesis, Department of Social Work, St. Louis University.

"Roster of the right wing in Illinois and Missouri" (1964) *Focus Midwest* 3, No. 6.

ROTHWELL, C. E. (1965) "Foreward." In D. Learner and H. Lasswell, Eds. *The Policy Sciences.* Stanford: Stanford University Press. (Reprinted)

ROUSSEAS, S. W. and J. FARGANIS (1964) "American politics and the end of ideology." Pp. 268–289 in I. L. Horowitz, Ed. *The New Sociology.* New York: Oxford University Press.

ROUSSEAU, J. J. (1762) *The Social Contract* (G. D. H. Cole, translator). London: J. M. Dent.

St. Louis Model City Agency and Human Development Corporation of Metropolitan St. Louis (n.d.) Metropolitan Housing Corporation Proposal.

SANFORD, R. K. (1966a) "Put up or shut up challenge helped start jobs project here." *St. Louis Post-Dispatch* (November 3).

————— (1966b) "Downtown St. Louis group gains in effort to preserve city's core." *St. Louis Post-Dispatch* (November 2).

————— (1966c) "Civic progress, inc. members contribute in team pattern." *St. Louis Post-Dispatch* (November 1).

————— (1966d) "Civic progress, inc. stays small, works through existing groups." *St. Louis Post-Dispatch* (October 31).

————— (1966e) "City aided by group of executives." *St. Louis Post-Dispatch* (October 30).

SAYRE, W. and H. KAUFMAN (1960) *Governing New York City.* New York: Russell Sage Foundation.

SCHMANDT, H. J., P. G. STEINBICKER and G. D. WENDEL (1961) *Metropolitan Reform.* New York: Holt, Rinehart & Winston.

SCHONDELMEYER, R., in "Religion and the great cities" (1967) Great Cities Series, St. Louis University (March 29).

SCHOTTLAND, C. I. (1963) "Federal planning for health and welfare." Pp. 97–120 in *Social Welfare Forum.* New York: Columbia University Press.

SCHULZE, R. O. (1958) "The role of economic dominants in community power structure." *American Sociological Review* (February): 3–9.

SEELEY, J. R. et al. (1957) *Community Chest; A Case Study in Philanthropy.* Toronto: University of Toronto Press.

SELZNICK, P. (1949) *TVA and the Grass Roots.* Berkeley and Los Angeles: University of California Press.

SHEPHERD, M. (1967) "McMillan re-elected head of St. Louis HDC board." *St. Louis Globe-Democrat* (February 25–26).

SILLS, D. L. (1957) *The Volunteers.* Glencoe: The Free Press.

SOLENDAR, S. (1963) "The challenge of social welfare." Pp. 3–24 in *The Social Welfare Forum, National Conference on Social Welfare.* New York: Columbia University Press.

SPEAR, D. H. (1949) "Economic composition of board of directors of private agencies belonging to the greater St. Louis Community Chest." Unpub. Master's thesis, School of Social Work, Washington University.

STANDARD AND POOR'S CORPORATION (1966) *Poor's Register of Corporations, Directors and Executives, U.S. and Canada.* New York: Standard and Poor's Corp.

STERN, P. M. (1965) *The Great Treasury Raid.* New York: Signet Books.

STRECK, P. (1965) "St. Louis mayoralty elections: the Negro and A. J. Cervantes." *Focus Midwest* 3:13.

SUTIN, P. (1966) "Gateway unit has new power to review U.S. aid requests." *St. Louis Post-Dispatch* (November 2): 1–2.

SWAYZE, C. (1967) "Businessmen back efforts for Pavilion fund drive." *St. Louis Post-Dispatch* (July 21).

TALMON, J. L. (1961) *The Origins of Totalitarian Democracy.* New York: Frederick Praeger.

THOMPSON, V. A. (1961) *Modern Organization.* New York: Alfred A. Knopf.

TIME MAGAZINE (1969) "Takeovers in high gear, threat or boon to U.S. business?" (March 7).

UNITED COMMUNITY FUNDS AND COUNCILS OF AMERICA (1965) "Guidelines for local participation UCFCA voluntarism study." Memorandum from United Community Funds and Councils of America, New York (August).

UNITED FUND OF GREATER ST. LOUIS, INC. (1967a) *Agency Budget Request for 1967, Organizational Services.*

——— (1967b) *Information.*

——— (1966–1967) *United Fund Handbook.* (1966–1967 edition.)

U.S. BUREAU OF THE CENSUS (1960a) *U.S. Census of Population: 1960,* Vol. 1 of the Population, Part 27, Missouri.

——— (1960b) *U.S. Censuses of Population and Housing: 1960.* Census Tracts. Final Report PHC(1)-131.

——— (1952) *County and City Data Book.*

——— (1940) *Sixteenth Census of the United States, 1940, Population,* Vol. 2: *Characteristics of Population,* Part 4.

U.S. HOUSE OF REPRESENTATIVES (1966) *Demonstration Cities Act.* Report No. 12341. 89th Cong., 2nd. Sess., (1966).

WARNER, S. B., JR., Ed. (1966) *Planning for a Nation of Cities.* Cambridge: M. I. T. Press.

WARNER, W. L., Ed. (1967) *The Emergent American Society, Large Scale Organizations.* Vol. 2. New Haven and London: Yale University Press.

WARREN, R. L., Ed. (1966) *Perspectives on the American Community.* Chicago: Rand McNally.

——— (1963) *The Community in America.* Chicago: Rand McNally.

WEBER, M. (1958) *The Protestant Ethic and the Spirit of Capitalism* (T. Parsons, translator). New York: Charles Scribner.

——— (1957) *The Theory of Social and Economic Organization* (A. M. Henderson and T. Parsons, translators). Glencoe: The Free Press.

WILDAVSKY, A. (1964) *Leadership in a Small Town.* Totowa: Bedminster Press.

WILENSKY, H. L. and C. N. LEBEAUX (1958) *Industrial Society and Social Welfare.* New York: Russell Sage Foundation.

WILLIAMS, R. M. (1952) *American Society: A Sociological Interpretation.* New York: Alfred A. Knopf.

WOO, W. F. (1968a) "Negro job problem obscured in confusion, discord, numbers." *St. Louis Post-Dispatch* (March 10).

——— (1968b) "Civic Progress, Inc. to leave HDC, set up own jobs plan." *St. Louis Post-Dispatch* (March 6).

WOOD, S. A. (1966) "United Fund drive was a team victory." *St. Louis Globe-Democrat* (November 12–13).

Yearbook of the Archdiocese of St. Louis, 1967 (1967) Issued by the Chancellery by His Eminence Cardinal Joseph B. Ritter, Archbishop of St. Louis.

ZALD, M. N. (1965) *Social Welfare Institutions.* New York: John Wiley.

INDEX

American Red Cross 92, 96
Area Progress Council 23, 40
Arts and Education Council 68, 75, 116, 184, 199

Bicentennial Civic Improvement Corporation 126-127
Blacks (Negroes), housing 161-162, 168-170; income 186-187; jobs 49-50, 172-180; newly mobilized 7, 31, 50-51, 142, 185, 190; politics 39, 45-46, 55-58, 85-86, 146-147, 171; population 37-38, 115, 122; services 103, 159-162, 172-179, 189-192
Boy Scouts 76-78, 103, 130, 138, 203
Business corporations, chief executives 29-30, 39, 64-65, 69-70, 75-81; locally based 21-22, 64, 69, 73; nonlocally based 63
Business, Young Men's Christian Association, and Urban League (BYU) 177-179, 184, 186

Calloway, Ernest 80-81, 83-87, 147
Catholic Charities 91, 116, 120-127, 132-133, 135, 204-205
Chamber of Commerce, Civic Center Redevelopment Corp. 149; fund drives 78, 91, 201; jobs 175; lines of power 28, 49, 198; Medicaid 183
Civic Center Redevelopment Corp., fund raising 182; limited profit 68, 101, 184; lines of power 150-151; politics 149, 156-158; tax free 150, 153-154
Civic Progress, general welfare 11, 37, 188; HDC 51-52, 186; interlocking directorates 70-73, 76; Juvenile Delinquency Control Project 49; lines of power 21-23, 28, 61-71, 87, 182-183, 191; Metro Survey 41-43; origin 38-40, 91; politics 55-58, 74-75, 78, 141-180, 197-199, 207-209; public housing 47-48; Racquet Club 68, 76; RIDC 49; sectarian influence 115-138, 201-202; UF, 89-113; see also BYU, CCRC, HWC, HPC, St. Louis University, Washington University, WOU
Civil rights, Birmingham 49; movement 47, 75, 198; populism 52, 55
Committee on Racial Equality (CORE) 46, 49, 160-161

222

Concentrated Employment Program (CEP) 174
Corporate Ethic 136-138, 144

Dahl, Robert 9, 19-21, 73, 184, 189-191
Democrats, business 83, 142, 190; mayorality 146; policy 167; voting 38
Downtown, Inc., lines of power 28, 74, 76, 149, 198; Mansion House 48; politics 75, 155

East-West Gateway Coordinating Council (EWGCC) 158, 166-167, 169, 179, 208-209
Elitism, brokers for 155-156; elite influences 21-22, 48, 77ff., 167-168, 178-180, 181ff.; large corporations 63ff.; metro reform 41-42; power elites 7-8; UF 89-113

Finpolity 25-26
Ford Foundation 39, 43, 46

Health and Welfare Council, agency review 49, 105-109; Cathedral Parish Study 127; employment 175-176, 179-180; Expenditure Study 123ff.; historical 91, 101; lines of power 28, 55, 68, 71, 73, 78-79, 89-113, 207; Resource-Needs Project 49, 102-104; source of information 9; see also HPC, HDC, Medicaid, MCF, MYC, Model Cities
Herbert Hoover Boy's Club 55, 78, 200
Horowitz, Irving Louis 7-9, 25, 142, 156, 187
Hospital Planning Commission (HPC), capital fund evaluation 125, 135-136; Civic Progress 74, 89, 104; finances 164-165; health planning 106-108, 166-167; HWC 55, 109; UF 68
Housing and Urban Development (HUD), EWGCC 158, 166-167; Model Cities 168, 172, 182-183, 203; Pruitt-Igoe 162; tensions 29
Human Development Corporation, see War on Poverty
Hunter, Floyd 9, 181-183, 188

Ideology 25, 32-33, 44, 189, 191

Jewish Employment & Vocational Service 115, 117-118, 138, 160-161